The Clash of Energy and Economics

OrangeBooks Publication

1st Floor, Rajhans Arcade, Mall Road, Kohka, Bhilai, Chhattisgarh 490020

Website:**www.orangebooks.in**

First Edition, 2024

ISBN: 978-93-5621-848-2

THE CLASH OF ENERGY AND ECONOMICS

NIKHAT KHAN

OrangeBooks Publication

www.orangebooks.in

I have dedicated this work to my only partner in my hard times, my sis Sarwat, and

To my life partner Amir!

About The Book

Generally books are meant to provide answers to our never ending questions; especially if the questions are based on our day to day life then it becomes a sensitive issue, and the answers are need to be very accurate without an error.

For solving the puzzle of economics, business, finance and the operating system I have written this book. It will help your business career as a journal, where you find all the pros and cons of past, present and future. I have used so simple words that a layman can also understand it well and practically apply to his life. Distributions of chapters are also so fine n clear that you will never get confused at any particular chapter at any particular time.

The book is divided into two parts; BOOK1 and BOOK2. The early chapters of BOOK1 will make your mind and built strong foundation for your thoughts on how the system of nature works', with the introductory note. The next chapters will address the energy, its transformation, future, role, and the connection with economics and world.

As the book move ahead readers will get the accountability of economics, its pre and post history, modern theories, current

Scenarios and further how they both (energy and economics) are inter - related to each other and how far they can go together.

There are four waves of energy and economics with four clashes of energy and economics has projected with complete data and future analysis.

BOOK1 has made the path for BOOK2;

BOOK2, is nothing but the Blueprint of this hi-tech century, in BOOK2, I will discuss the new economic thought which needs to be implemented on Artificial Intelligence era, opportunities for SME's, investors and for those who wants to take the full fledge advantages of this upcoming industrial revolution and make their names on the pages of history.

Contents

About The Book vi

Contents vii

The Opening 1

    Alchemy 2

1. ENERGY 5

    The Pattern of Energy 7

    The Transformation of Energy 10

    Conversation of Energy 10

    One Coin, Two Sides 13

    Time Machine 14

2. ECONOMICS 21

    Journey Of Economics 23

    Economic Cycle 24

    Pre-Classical 25

    Mercantilism 27

    Commercial Capitalism 28

    Classical Economy 36

    Marxists' Economy 46

    First Wave of Energy and Economics 50

    Neo – Classical Thought 54

    Alfred Marshall 54

    Second Wave of Energy and Economics 60

    Modern Economic Thought 64

    Third Wave of Energy and Economics 76

3. THE CLASH 80

    The first clash - Britain v/s Dutch 82

    The Second Clash - Britain v/s colonies 85

    Third Clash - The Age of War 90

    The Third Chain Reaction 91

Summary of Book 2 95

BOOK – 2 .... **97**

THE 4ᵀᴴ QUADRANT **98**

The Fourth Quadrant 100
The Fourth Wave Of Energy And Economics 102
The Fourth Clash; Catastrophic Clash 107
The Global Economic Thought 110
The Global Economic Machine 114

Summary of Book 2 .... **134**

References **135**

The Opening

"The beginning of knowledge is the discovery of something we do not understand".

- Frank Herbert.

With the advent of Big Bang everything has came into existence, all living organisms, species, planets, several creatures, they all spread to their specific place or regions for specific time at the level of certain energy which was already inbuilt in their system for functioning accurately. The celestial sphere has covered the earth by ozone layer for protection; planets get placed in their orbits and operating continuously without fail, a long series of day and night with several seasons appeared, sometimes its rain which come and flourish the earth, sometimes its heat & summer to balance the environment, then comes winter & spring which flow old leaves to make new ones.

But why this celestial and ecosystem are working? To whom they are providing so much of services, with their benefits?

The stage of world is prepared, organized, developed, synchronize and operate only for human's survival, and to make our survival easier & beneficial on earth we have one power, if we use that power in proper way nothing will going to stop any person or community or the society.

If we go through the history of civilization, we found that one thing in common to each era, their rise or decline, is the use of that power, they had found the alchemy of this power to be in power.

Alchemy

The world is full of matters, when humans started their journey on earth; they got introduced with so many new factors, objects, and the law. The matters around us are there for a reason, they are one kind of raw material which can transform into energy.

This knowledge takes the Homo sapiens to another level of innovations. Although it was Albert Einstein who gave this world changing formula;

E=MC2

But the use of this formula was found in prehistoric period, where the – superpowers of that time, like; Babylon, Rome and Egypt were making their Empires and societies.

There are three principles of nature which lead any nation, state or community towards rise and maintain them at peak, if any of these lack, the decline will happen spontaneously. In early civilization we found footprints of this universal law, by practicing them, Babylonians, Egyptians, Romans, Chinese, Iranians, currently Europeans, Russians, Israelis and the United States has maintain their balance of power.

The Universal Law;

1] Knowledge And Morality;

Knowledge is the key of human's survival and if get combined with morality it takes humanity to heavens. All the successful people who make changes in the world are full of knowledge with wisdom, along with knowledge they applied it on different platforms to generate results, there moral values for their knowledge and work will never get compromised in anyways. They understand the pattern of energy which is working in this planet. Each and everything around us have some sort of energy which always got in the process of conversion, and make another form of energy. They use this law of conversion of energy by understanding the mass, light and time theory. In this era also where technology is at its peak, various of innovations are taking place which never had happened previously, world has become so competitive which never had been before, so in that meantime if we also want to make our position stronger than ever, then the formula is simple, we also had to acquire & apply the law of conversion of energy for innovation, but not just the energy, the energy always comes with economics.

2] Perception And Optimism;

When the individuals of any societies have clearly defined their principles, their mind full point of view for every situation torch good outcomes, with their perception they set agreements, and by their optimistic approach, they make alliances & set forth the constitutions for the world. Most current example of this we had witness in 1993 by European Union, how the European communities and their individuals with their clear mindset had dictate their formation along with its rules & regulations, and make firm bricks of socio-economical-political structure.

3] Political Status;

None of the organizations or communities will come upfront if they don't gain their political values. Without politics everything incompletes, and with politics everything work in their order. The political status around the world of any nation or community has decided their category on world platform. Either they are independent, autonomous, developed, developing or just colonies. When you came on world platform with visionary knowledge combining morality, perception & optimistic approach, and most importantly the socio – economic pillar, these ingredients will allow the political campaign for leadership of any nation.

These three universal laws had applied by our ancestors, early super powers, and unbeatable empire's. So, we have come to the conclusion, that energy is the elixir, and there is strong connection between energy and economics. With this knowledge everything is possible but without its literacy, we will definitely go astray. All the big events of the world, revolutions, Renaissance, or the rise & fall of nations, energy-economics are the major factor. The understanding of this pattern will lead you to success in any field, whether its business or life, every sector required it, but without knowing the pattern of energy-economics relation many start-ups, or sometime giant organizations too fail, and crises or depression occurs. So at very first we have to understand this illuminious pattern of energy and economics.

But before starting it, I have one question for you, what if we are out of energy? What will happen to the world, if there is no energy? Will economics also survive without it?

Answers you get in further reading of this book. But read it in analytical way; keep you mind and thoughts open, so that more questions can arise. As I already mentioned that this book is nothing but a manual for those who want to make their position in coming times of new world and geopolitics.

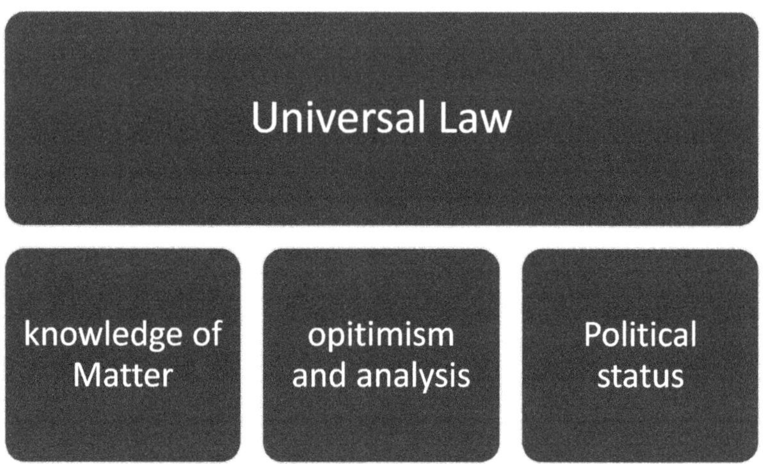

(Fig, 1; showcasing the pattern of Alchemy)

1. ENERGY

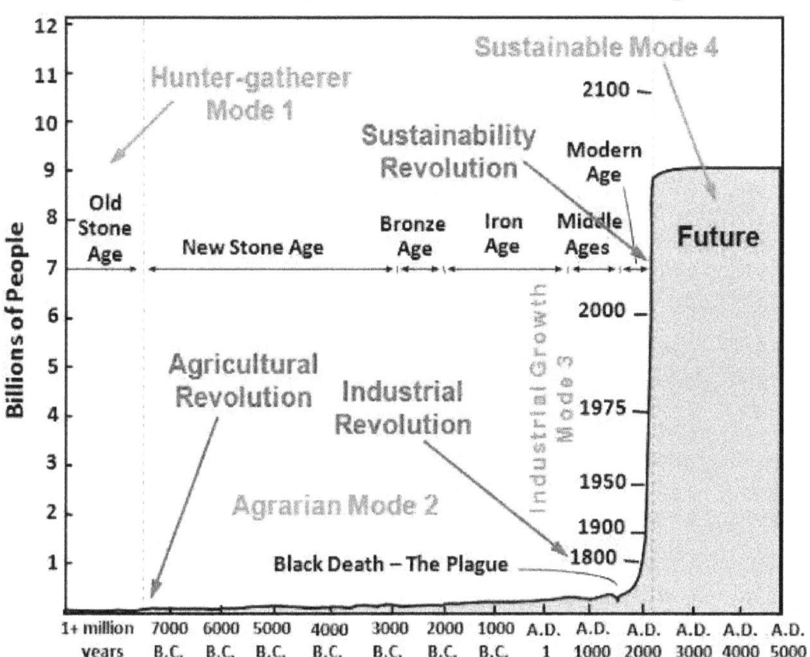

(Fig, 2; explain the transformation of energy)

This graph is represented by United Nation, world Population summary and modes of human history. But these four modes of human history also draw parallel of energy history and its evolution.

Since the beginning, energy is present everywhere, but as you know, humans take time to understand things, matters, application's and consequences. Prehistoric civilization torch hunter-gatherers at Stone Age, with very less use of energy because it is the first phase of human development. Then, it come the fire, and with its sparkle humanity gets illuminated to change, fire, is the first component by which change in human life has started and conversion of energy too.

Man has started walking to the path of knowledge and practice, with the time, awareness has waking up, awareness of self, protection, food shelter, garments, crops, soil, climate, nature, animals, resources, heaven and the earth, their link and the linkage between all living things present around, because of that much awareness, political and economic occurrence seen in the society at ground level, but it dominate the whole system with the coming ages.

Bronze and Iron Age, introduces the energy and chemical elements to world, which again helps in promotion of industrial revolution, whereas the industrial revolution, changed the concept of energy itself in so innovative form that, no one can distinguish between renewable and non-renewable, lives become so dependent on non-renewable, that human minds are in continuous thinking of more and more conversions in the field of energy.

Yet the new world order began, with new geopolitics, economics and social life. Things become more digital and technology is at its peak. And in this highly technological era, demand of consumption of energy has increased vigorously. Perhaps, change makers are engaged in providing artificial intelligence to the world, our only focus to AI, may be advantageous but somewhere contagious also.

Some visionary reformist has also analysis that the future of energy will be sustainable again, not completely how it had been started, but in developed primary form. Also, it's the law of nature, called; Eternal recurrence[1], which denotes that, Time is a loop and this in loop, the exact same events will continue to occur in same way as happened earlier, this loop from first to last point get reverse by last to first again, for eternity. This prediction of sustainable future has shown in the diagram of UN also (mentioned earlier, fig1).

This can understand by nature also, as everything had started because of big bang theory, but before this event, there is nothingness, just a total blackout. Again in loop, nothingness and blackout will appear to world after the reckoning event of Big Crunch.

Although, the conversion of energy has need to be understand first for our successful journey ahead. Moreover, it easily gets comprehend by its pattern; there is three phases to understand this pattern of conversion;

1. Primary phase of energy,

2. First phase of energy,

3. Double Transformation.

The Pattern of Energy

"Things are always at their best in their beginning".

_ Blaise Pascal.

Human civilization had started with clean energy, which normally we called as Renewable energy. Those renewable resources such as [i.e.; solar without panels; biomass without laboratory procedures; wind without turbines; hydropower without dams; and geothermal without cable wires] are the only things on which early man has so much dependent for his living. The early man, who usually called hunter-gatherer in modern terms are the prime users of sustainable resources with the very simple living standards. After hunter gathering, there comes a time, where people realizes the power of energy, the power of sunlight, winds, and water by combining all, the first revolution of human history occurred, "The Agricultural Revolution".

Whereas Agricultural Revolution in Egypt, France, Greek and some minor continents of Europe had tasted harvesting through solely clean, primary and sustainable energy. Along with sustainability, this universal energy also employed all people of the society from various of age factors for development.

(Fig, 3; sketching clean energy of prehistory)

Since the agricultural revolution, sustainable energies was in high demand, societies had started moving for betterment, people had started experimenting these energies by different methods, for fruitful harvesting some great minds invented canals for proper water supply, these canals were hydropower plant for that time. Cattle also played big role in agricultural revolution, apart from heat, air, water, canals and cattle's, there is one more thing, which is the key player, not only for this revolution but for coming event of industrial revolution, and that key player is; Iron. Sharp weapons, hard equipment's for farming all made from iron. It also use for self-defense and in wars, with the advancement of human mind, the era has successfully reached to 1800.

{Before 1800, energy has used in its primary form, there is no great conversation, it only extract the raw mass of energy and make it in the pure form. So we labeled this energy before 1800; - (E1).}

At the dawn of 1800, world get introduced by some different kind of energies, which has never seen or used before. From this point a series of innovations had started, tremendous forms of non-renewable energy and its transformation takes place, with multi-dimensional uses, and this birth of non-renewable energy was integrated with respect to time.

Steam engine has the major role in changing world and bringing more upcoming revolutions, because of steam,; railroads, motor boats, runs the economy. Afterwards iron and steel, coal, fossil fuels, natural gas and mother of all, atomic energy, witnessed by people.

{At this point, transformation of energy occurred, by the primary energy, it also called, first phase of new energy, this denotes with (E2). And the equation becomes like, E1 $\longrightarrow$ E2.}

Perhaps, with the use of this energy, the standard of living had increased in big volume. These energies has taken the human race to the modern world, and in this modern era, where future is discussing by thinkers, the competition of technology and power is making a path for the fourth industrial revolution, which is also claiming the mother of all revolution like atomic innovation in energy sector.

{The double transformation of energy denotes with (E3), combined by primary energy and first phase energy & the equation will now seems like;

E1 $\longrightarrow$ E2

E1 + E2 $\longrightarrow$ E3

=

E2 $\longrightarrow$ (E2 – E3)

= -E3.

The end result is (-E3), which means that the energy which will take in charge of coming times, world economic, social, political structure, global and digital world, is the artificial intelligence of fourth industrial revolution, which no one can stop, it is proved by this equation. But just one thing is remarkable here, is negative sign of E3, is it will run in minus, or some negative factor? What is that (−E3)? This minus concept will resolve in the chapter of The Clash.}

Primary Energy _ **E1**

First phase of energy _ **E2, (E1 + E2);**

Second phase of Energy _ **-E3**

The Transformation of Energy

"Energy cannot be created or destroyed; it can only be changed from one form to another".

- Albert Einstein.

Why we need to learn the transformation, how it will benefit our present and future? We will find answers of these two questions, along with analysis.

Financial Independency is the common desire in all human being, for financial stability some people do labor, some of them hunt jobs and some creates both job and labor for people by organization, called; business, which make an environment of engagement by people, place and market. Here, we required a lot of energies in various forms, but have you noticed one thing, we required energy for businesses and organizations which literally means an economy, meanwhile, energy also boom with economics.

Conclusion is energy and economics both are inter-related, and inter-dependent on each other, it's like two sides of one coin, if one side you have energy then on the other side you definitely have economics only.

We will draw a synopsis, from history to present world in which we found both energy and economics working together and how their inter-connected relation make changes throughout century. And for

This century also their dynamics are going to work similarly, this is the reason of learning, its transformation process, because the transformation of energy is transformation of economics only. And the ignorance of energy is the ignorance of economics, which we can't afford.

Conversation of Energy

After the very first moment of Big Bang, the surface of earth gets covered with so many energy, various of metals and stones and different kinds of chemical elements. From this point, conversion and transformation process begins for energy, some occurred simultaneously, or some required step by step procedures. Perhaps, the mother of all energy, which makes the path for upcoming revolutions, was already present on earth thanks to Big Bang, namely:

1. Gravitational energy

2. Kinetic energy

3. Chemical energy

4. Potential energy

Basically, these four magnificent energy was solely responsible for all the innovations.

[1] The law of conservation of energy states that the total energy of an isolated system remains constant; it is said to be conserved over time. In the case of a closed system the principle says that the total amount of energy within the system can only be changed through energy entering or leaving the system.

However, there is relativity on mass-energy equivalence, which mass need to be applying with respect to how much particular of time/light, will going to transfer energy from one form to another. Fire, is the best and first foremost example of this transformation process, where friction was involved in creating that, whereas friction is nothing but the kinetic energy which produces heat, this heat sparkle the light and the combination of both called fire, which is again an energy called chemical energy. i.e.

Kinetic energy $\rightarrow$ heat + light

Heat + light $\rightarrow$ fire

Fire $\rightarrow$ chemical energy.

{Energy transforms into another form of energy}.

After fire, other conversions also had happened in the same manner, i.e.

- Gravitational potential $\rightarrow$ heat

- Heat $\rightarrow$ thermal energy.

- Chemical energy $\rightarrow$ mechanical energy.

- Wind/wave power $\rightarrow$ mechanical or electrical energy.

- Heat + electrical energy $\rightarrow$ geothermal energy.

- Heat + mechanical energy $\rightarrow$ steam engine.

This chain reaction finally brings the age of first industrial revolution with the innovation of steam engine by heat and mechanical energy combustion process. And again it is in continuous form till in the age of fourth industrial revolution, but the reactions are happening only in mechanical and electrical energy now. Which we already had seen in previous chapter [The pattern of energy], with the result { -E }.

This transformation had created one more transformations, which is economic transformation, operating parallel to energy.

Energy ‖ Economics

One Coin, Two Sides

ENERGY	ECONOMCS
Stone	Hunter-gathering.
Bronze, iron	Barter trading.
Fire	Food, security and traveling.
Water, wind, fire	Agriculture.
Wood	Shipping.
Forest, mountains and nature	Poultry, cattle's, mining and natural resources.
Rock, metals	Gold, silver.
Steam engines	Manufacturing, transportation.
Steel and coal	Railroads, factories and vessels.
Electric energy	Electricity, networking, cables, internet and digitalization.
Artificial Intelligence	Artificial Economy.

This table explains everything about both energy and economics, and their conversions with developments. From the cave to skyscraper, from stone to computers, from herdsmen to machine learning, from gold mining to data mining, from wheat to hamburger, from raw cotton to brands and from basking to nuclear explosions, this all possibly happened because of energy and economics together, whenever they work in each specific direction, the outcome has shocked the world again and again.

Present Man has come so far that he can't even imagine or think the evolution of his manhood, lifestyle and his beliefs. The world from last 250 years have changed so fast and spontaneous that we forgotten that once we had lived in rural or below than rural. But in this breathless changing system, if we leave our roots and lessons from the past, our future will then get unsecured and in complete disaster. In fact, not just ours but our generation's, imagine your child, or grandchild after 20 or 30 years from now, and you are telling them about your ancestor's and their living, they will be gone a laugh, cause at that time, the system and technology will work on some another level that they can't make them realize the old scenarios. How will you tell them, that we are in titanic heading towards iceberg, only thing which will help ourselves and our generation is The Time machine.

A Time machine in which we can see all the things of past, dos & don't, strategize the consequences, apply information's to future without be mistaken.

"From this point of chapter till The Clash, read all the material by the lens of Time machine only".

Time Machine

[2] *The first Industrial Revolution started in Britain's textile industry in the mid – 18 century, sparked by the mechanization of spinning and weaving. Over the subsequent 100 years, it transformed every existing industry and gave birth to many more, from machine tools to steel manufacturing, the steam engine and railways. New technologies led to shifts in cooperation and competition that, in turn, created entirely new systems of value production, exchange and distribution, and upended sector from agriculture to manufacturing, from communications to transport. Indeed, the way we use the word "industrial".*

From this first wave of transformed energy, the environmental changes started appearing, it is not just shift of energies, but a big shift on geographical pattern. The birth of revolution had come into existence at a place where population was not much efficient to cherish growth of industry and form second wave. Thankfully, we had discovered steam first; it has helped British East India Company to make the colonies, gain huge investments, combined skilled workers of different regions at very low payroll, and maximize their outcome.

Meanwhile, it creates big difference, difference for some countries like Britain itself along with France, Netherland and some other European countries get wealthier instantly, not just Europe making mountains of Gold, but surprisingly, the newly established country like America growing its per capita income annually. Whereas for some countries it's just matter of food, cloth and shelter, they are not working on per-capita but for per-day income.

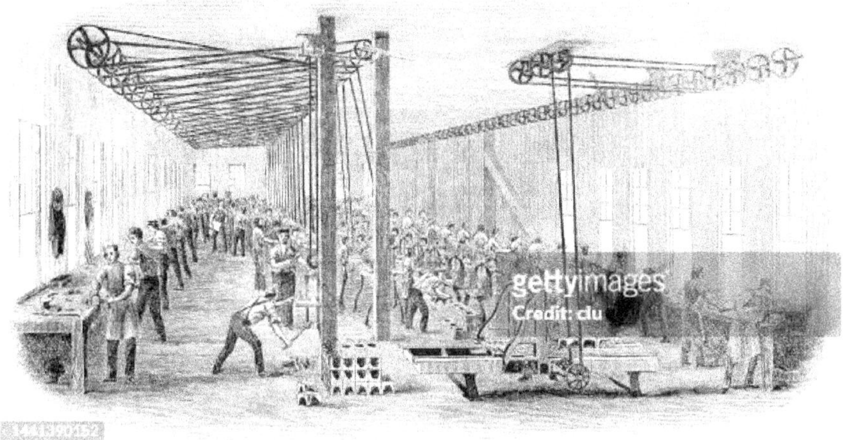

(Fig, 4; workers in textile mills, steel and coal mining, loading goods for shipping and transportation labor for per day wages_ this is the picture of our Past).

[3]*In the period between 1870 and 1930, a new wave of interrelated technologies compounded the growth and opportunity that came from the first Industrial Revolution. The Radio, telephone, television, home appliances and electric lighting demonstrated the transformative power of electricity. The internal combustion engine enabled the automobile, the airplane and ultimately their ecosystems- including manufacturing jobs and highway infrastructure. There were breakthroughs in chemistry: the world got new materials, such as thermoset plastics, and new processes – the Haber-Bosch process, synthesizing ammonia, paved the way for cheap nitrogen fertilizer, the "green revolution" of the 1950s and the subsequent spike in human population. from sanitation to international air travel, the Second Industrial Revolution ushered in the modern world.*

The power of steel we witness here, at that people were amazed that how can a piece of steel, with the combination of other metals like, titanium, aluminum and other alloys can flying and not just flying but carrying 1000s of tons. In present scenario, it is not a very big thing to discuss, when some people have their personal choppers, air tanks and air based military. The first person who lit the fire from two stone by rubbing them together had no idea that this form of fire got converted into combustion process in future and with this fire so many highly defined things will happen.

(fig, 5; shows the outcome of second industrial revolution)

[4]*Around 1950, revolutionary breakthroughs occurred in information theory and digital computing, the technologies at the heart of the third Industrial Revolution. As with the previous periods, the third Industrial Revolution was not due to the existence of digital technologies, but to the ways in which they changed the structure of our economic and social systems. The ability to store, process and transmit information in digital form reformatted almost every industry, and dramatically changed the working and social lives of billions of people. The cumulative impact of these three industrial revolutions has been an incredible increase in wealth and opportunity – at least for those in advanced economies.*

[5]*Illustrative contributions of Industrial Revolutions to Humans Development: OECD Countries 1750 – 2017.*

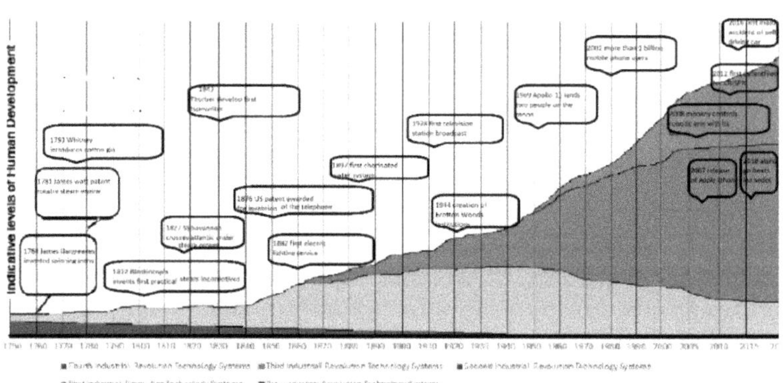

[6]*The figure shows that, even for countries close to the technological frontier, the lion's share of human development comes from the technologies and systems developed during the second Industrial*

Revolution – such as electricity, water and sanitation, modern healthcare and the huge expansion in agricultural productivity driven by the invention of artificial fertilizer. This is an argument that Robert Gordon and others have made persuasively.

But the interesting fact here is that, OECD previously known as OEEC, has established in 1948, perhaps the human development was particular at Europe, and for remaining countries, human development was below average. It clearly indicates that it is not OECD who promotes economic growth of its Alliance countries but it was the already present wealth among them, who firstly create OECD and then started operating under that pre-determined charter.

It is not idle, if we say, inequalities has also risen with Industrial Revolution only, gap between rich and poor get expand, the circulation of wealth also get in limited hands, but this is not happened with magic, there is chronology of industrial revolution involved, for making some country and people rich and richer, and to some country and people poor and poorer. (Discuss ahead) Along with chronology there are geographical benefits, climate benefits and easy commute of transportation.

On a philosophical point of view, nature also helps those, who work hard for humanity and for easiness of humanity; our intensions are the most countable thing in this universe for working either along or against us. People from the past, i.e.; philosophers, thinkers, scientists and writers, they all contributed their hands to this world changing revolution and because of their efforts they got responded.

[7]*The industrial revolution has surged forward in three major chronological phases—the first, when it began and spread directly only within the West; the second, when it matured and began to exceed Western boundaries; and the third, when it became effectively global. Our lives today are being shaped by this ongoing third phase. And in each phase, industrial chronology cut across more conventional historical divisions, such as the French Revolution or World War I. Industrial history has its own chronology, and it must be seen as an ongoing process.*

Sun always rise from East, but the sun of Industries has raised from West, along with British, United States, the newly formed nation had entered into the historically successful economic and financial journey. And this huge success lead U.S to became the sole supreme power of unipolar world on world map. Slow expansion seen after decolonization. For competing U.S, Soviet Union had been came forward with the rise of Russia in industrialization, it had again turned the world map from unipolar to bipolar world. After Russians, East had tasted the flavor of industries, bipolar world also vanish with the fall of Soviet Union and world has taken new shape of new world order with multipolar world. Where we saw several blocks of political economy, whether it's NATO, G20, EU, or GCC, Asian-pacific block, OIC etc.

[8]*World history from the late nineteenth century to the mid-twentieth: industrialization outside the West, redefinition of the West's industrial economy, and growing involvement of nonindustrial parts of the world, along with an overall intensification of international impact.*

Russia's industrial revolution had a massive impact on world diplomacy. Japan's revolution altered world diplomacy as well and ultimately had an even greater effect on the international balance of economic power.

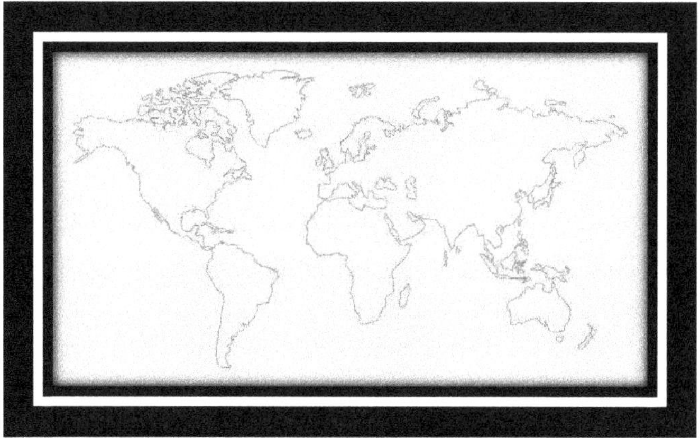

(fig, 6; beginning of industrial revolution from Europe)

- Location of Britain connected Europe and U.S, surrounded by Atlantic Ocean, North Sea and English Channel – unipolar world,

- The cape route, which is used by British East India Company for Imperialism.

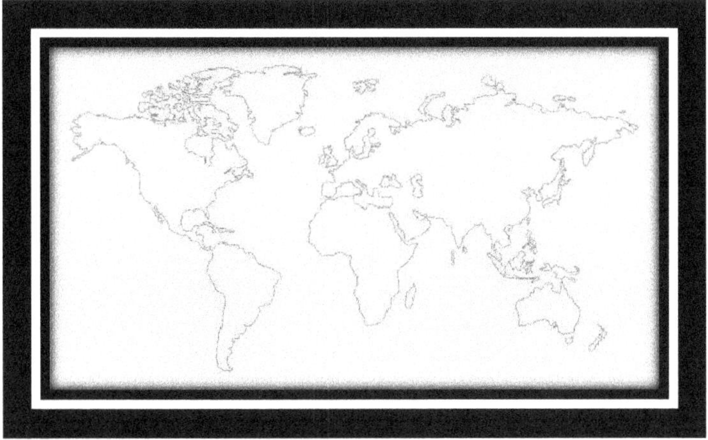

(fig, 7; second phase world power emerging from Russia.)

- Its geographical condition helps communism to rise,

- Russia influenced far east countries,

- China and Japan had major influence of Russia,

- Because of Russia, industrial revolution gets spread in Asia and globally, Russia work as bridge.

(fig, 8; third phase of multipolar world blocks)

[9] *'The world's third phase of the industrial revolution had several primary facets. First, there were a number of major new industrial revolutions. Expansion started slowly, with particular focus on the Pacific Rim. But by the 1990s a number of huge economies, including India and China, pulled*

into the process. The revolution that had begun two hundred years earlier in Britain had not played itself out yet. By the twenty-first century over half the world was effectively industrial for the first time. Regional inequality remained an agonizing problem, but its dimensions were redefined. Second, at the same time, established industrial societies moved toward a new set of technologies that had substantial social implications. Some observers talked of a third, or postindustrial, revolution in trying to convey the magnitude of these new developments. Continued change in advanced industrial societies, plus the surge of newcomers, raised vital questions about mutual relationships, such as, how could old and new industrial economies best interrelate? This facet was related strongly to the preceding categories of change—industrialization had a more decisive impact on the international framework than ever before. Communications accelerated, commercial contacts moved to new levels, and industrial units operated worldwide in a process that came to be called "globalization." The industrial revolution, which had already changed the nature and extent of international contacts, now burst beyond the bounds of nations and even whole civilizations. Finally, global industrial growth helped to generate a new level of social and environmental change, altering many aspects of the human experience in most of the world's regions.

2. ECONOMICS

There is no presence of word "Economics" before the first industrial revolution, or before 1776. Perhaps, there is existence of economics but not in that defined form, which is majorly used in current world since the discovery of factories. Although the needs and wants of humankind is not modern concept, it had started at the day, when journey of human life began on the earth; Man's survival is the topmost thing. Whatever is the purpose of human life would be but this is a proven fact that from the day of his creation, his first effort has always to fulfill his primary needs. Whether, need is basic for all, though, when wants come, which is a form of desire, here imbalance of society occur, and journey of economics start.

Economics is the byproduct of wants, which is the reason; it had emerged during first industrial revolution, where need of humans shifting to desire. It was his need which forced him for hunting, harvesting, hand crafting etc. whereas afterwards it changed into his desire which make him leave rural areas and form urbanization, sewing to textile mills, and hand crafting to machine crafting, this manufacturing units and industries needs an operation which perform by economists.

After desire then comes the battle of wants, the unlimited wants of people generate unlimited demands, these unlimited demands transfer into unlimited supply chain to match the demand; perhaps it's not an easy task. It required a bulk of organizing units, who will perform all the activities without fail, Technology get born for fulfilling that purpose.

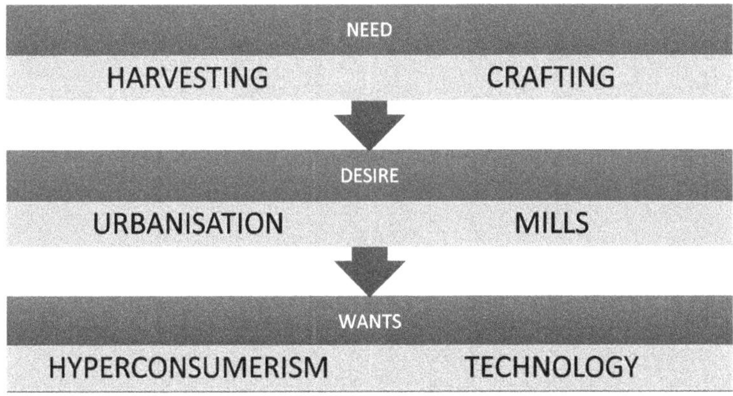

(fig, 9; changing phases of economics)

We can summarize it in a single phrase, "world has divided into two parts; world before economics, and world after economics". Before economics, it was nothing but a trap called; The Malthusian Trap, it named after Thomas Malthus, he has discussed and given a graph which discriminates economics before and after 1800, also summarizes population growth.

After Malthusian trap, there is emergence of economics with collaboration of industrial revolution, surprisingly; the income of world population which was floating uniformly got raised highly but on limited countries followed by divergence.

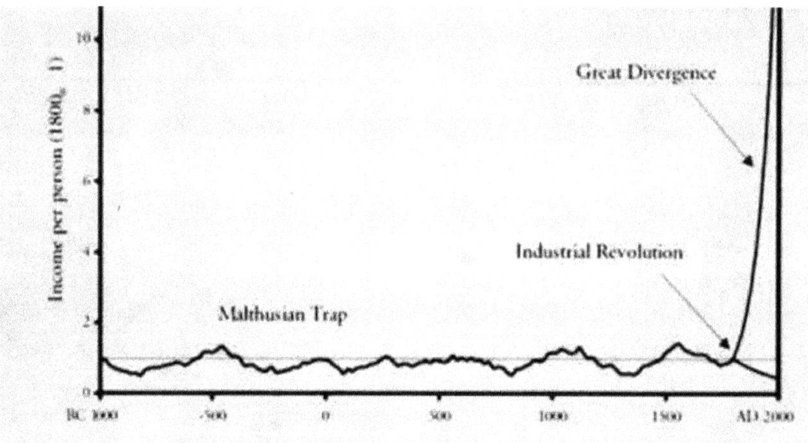

Figure 1.1   World economic history in one picture. Incomes rose sharply in many countries after 1800 but declined in others.

(fig, 10; Malthusian trap, essay on the principle of population-1798).

World after economics also witnessing inequality since 1800, when income started declining in other countries where making some countries highly wealthy. The circulation of wealth had begun in European continents, under the flag of Dutch; this flag had passed to British after Industrial rise, with the rise in population too as we can see in Malthusian graph.

Population has seen increasing globally, but the income was increasing specifically to some regions of industries, the remaining's of populated regions that don't have incomes, they have slavery for those who had it.

For managing these trials, some of the intellectuals of societies joined on the table together to create a charter for wellbeing of people, society, economics and politics, and from here begun the journey.

Journey Of Economics

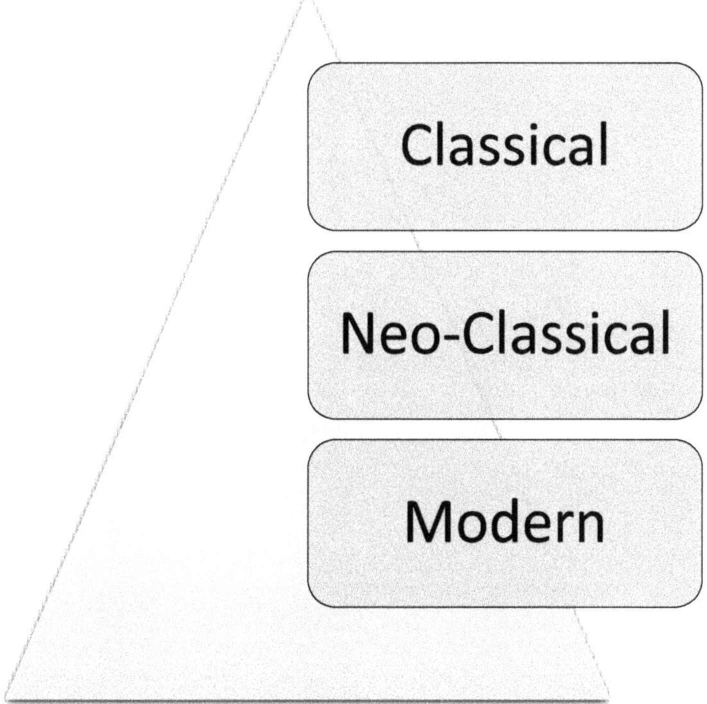

(fig, 11; Economic foundation, where Triangle represents pre-classical economic Era).

In this age of 21st-century, the field of economics had been already passed through three phases: classical, neo-classical and modern thought. Currently we are applying modern dynamics to changing world order, but these dynamics were not renewed for this and coming generation, apart from these three, the building blocks of economy has inserted on pre classical economic thoughts only, which represents in triangle(fig). This triangle is actual structure of economics and society in where three thoughts have inserted for regulation. Explanation of this structure will discuss in pre classical economic thought.

Economic Cycle

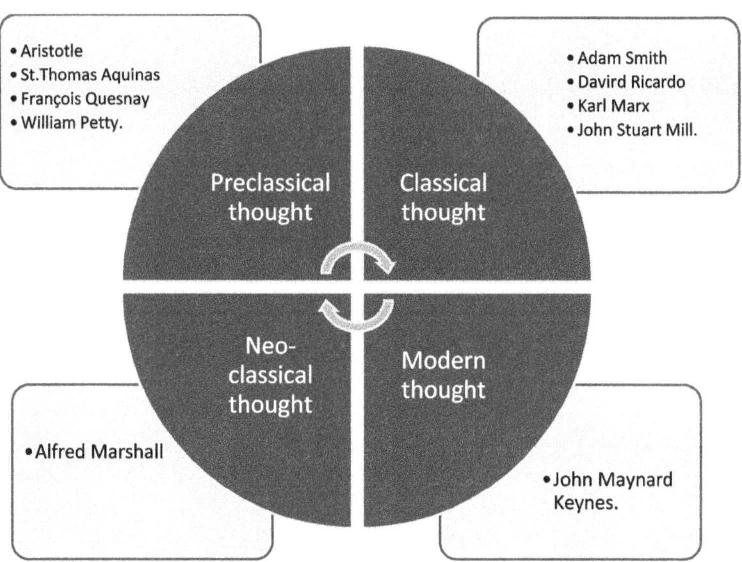

(fig, 12; Representing four phases of economic thoughts and their contributors).

Why we are using the term cycle is very simple, economy has always works in circular motion in their orbit, and the orbit of economy is market. Hidden forces of market play snake and ladder with economy,

(Circle – economic cycle, and rectangular – hidden market forces)

it goes rise, firm, stationary, but then all of sudden it gets down again to point 1, from where it had started, basically; market and its forces are shaping any economy.

Pre-Classical

Scarce and allocation of resources are two critical questions from early times till now. Some philosophers tried to solve this puzzle by their different approach, in their result, Ancient Egypt and Babylon had gained the power with their planned economy of irrigation system, the Assyrian and Babylonian theocracies had built huge military & bureaucratic establishments and elaborated their legal systems of which the code of Hammurabi, (is the earliest legislative monument about 2000 B.C). In the east; Kautilya in India, and Guanzi in China had solving their economic issues;

[10] *In India Kautliya's Arthasastra, dealing entirely with the functioning of the state in its economic aspects, belongs to the fourth century BCE and is full of references to previous texts, In Chaina, Guanzi brought together writings dating from the fifth century BCE and the first century CE dealing with a variety of matters including economic issues.*

Whereas in the west; exchange of goods on the medium of money, already existed in Athens, Greece, and Caesar's Rome. They had already pursued activist foreign policy; also they had monetary institutions to a high degree of perfection and they had expertise in credit and banking.

Markets and transportation of goods by Caravan Trading was the earliest form of economic and social development, through which Man becomes more civilized and globalized.

Meanwhile, Aristotle the great philosopher came forward with his ideas and from here, physiocratic approach to society had started.

[11]*Aristotle's main contributions to economic thinking concerned the exchange of commodities and the use of money in this exchange. People's needs, he said, are moderate, but people's desires are limitless. Hence the production of commodities to satisfy needs was right and natural, whereas the production of goods in an attempt to satisfy unlimited desires was unnatural. Aristotle conceded that when goods are produced to be sold in a market, it can be difficult to determine if this activity is satisfying needs or inordinate desires; but he assumed that if a market exchange is in the form of barter, it is made to satisfy natural needs and no economic gain is intended. Using the medium of money, however, suggests that the objective of the exchange is monetary gain, which Aristotle condemned.*

According to him moral conducts are more important than anything else, his thoughts are very ethical towards socio-economic pattern and politically also he was so religious on natural laws which laid the foundation of antiquity and later the Ecclesiastical polity.

After the physiocracy, Western Europe had adapted Scholastic approach towards economics during thirteen century, where Thomas Aquinas' arguing Aristotelian thoughts.

[12]*Aquinas and other scholastics were also concerned with another aspect of greater economic activity, the price of goods. Unlike modern economists, they were not trying to analyze the formation of prices in an economy or to understand the role that prices play in the allocation of scarce resources. They focused on the ethical aspect of prices, raising issues of equity and justice. Did religious doctrine forbid merchants to sell goods for more than they paid for them? We're making profits and taking interest sinful acts? In discussing these issues, Aquinas combined religious thinking with Aristotle's views. When exchanges take place in the market to meet the needs of the trading parties (using Aristotle's conception of need), Aquinas concluded, no ethical issues are involved. But when individuals produce for the market in anticipation of gain, they are acting virtuously only if their motives are charitable and their prices are just. If the merchant intends to use any profits for self-support, for charity, or to contribute to the public well-being, and if his prices are just, so that both the buyer and the seller benefit, the merchant has acted rightly.*

Merchants created boom in economics, society needs statistics of their goods and trading with accountability, this invented an economic term which fuels the economy and it termed as National Income, because of merchants trading, goods needed to be recorded, measured and answerable to society, for performing this work William petty with his crude method entered in the world of economics. Petty with his static technique tried to measure and solve the growth of society, population, their income rates, import and export of goods and the nation's capital stock.

These all are the questions of economics which lead Adam Smith to answer.

After Petty, in the year of 1694-1774, the influencer of Adam Smith with his different view on economics and his calculative economic table enlighten the world by providing new aspects for economy and society, his name was Francois Quesnay.

[13]*The physiocrats considered Quesnay's "economic table" to be their crowning theoretical achievement. It gave a crude representation of (1)*

the flow of money incomes between the various sectors of the economy and (2) the creation and annual circulation of the net product throughout the economy. Quesnay's table represents a major methodological advance in the development of economics—a grand attempt to analyze raw reality by means of abstraction.

Just two year after Quesnay, Adam smith had introduced "The wealth of nations" to the world in 1776, and became medalist for the father of economics. Quesnay had major role in Smithians thought, along with his contribution for Classical economic development.

(fig, 13; evolution of economic thought).

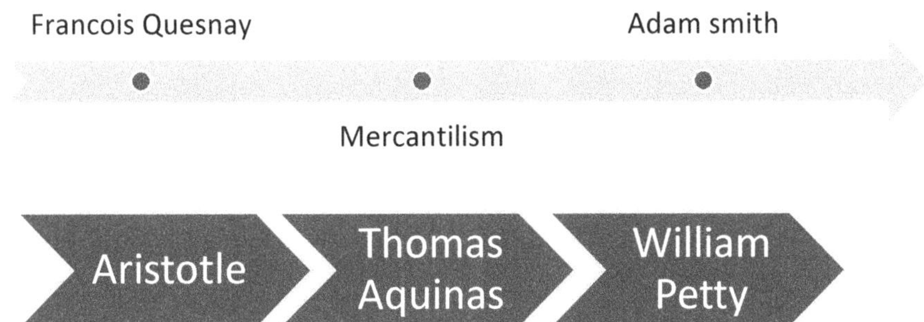

Before classical phase of economics, one more event had taken place on the platform of economic development, which given building blocks to capitalism, economists named it; Mercantilism, the first brick on the skyscraper of capitalism.

Mercantilism

[14]*The mercantilists proceeded on the assumption that the total wealth of the world was fixed. Using the same assumption, the scholastics had reasoned that when trade took place between individuals, the gain of one was necessarily the loss of another. The mercantilists applied this reasoning to trade between nations, concluding that any increase in the wealth and economic power of one nation occurred at the expense of other nations. Thus, the mercantilists emphasized international trade as a means of increasing the wealth and power of a nation and, in particular, focused on the balance of trade between nations.*

This thought had given fire for industrial revolution, also three factors of economics had generated by mercantilism, namely; land, labor and capital. During this period export had risen unexpectedly, leaving import behind, feudal society had been formed for individuals and organizations but mainly church had seen with all the major control. People who were living simple form of life with their income, started holdings, surplus created here; which again demands the price and value of money. Caravan tradings shifted to cargo trading with transportaion revolution too.

The science of mercantilism is the most important aspect of economy, without having the understanding of mercantilism, we can't understand economics itself. This science had briefly explained by one of the best historian of 20'th century, Caroll Quigley.

Caroll in his renowned book 'Tragedy and Hope', sketched the complete history of rise and fall of human civilization specially in economic term, while discussing about mercantilism, he labelled it not with mercantile movement but with commercial capitalism, which again rooted from western countries, led the foundation and firm pillar for many things; capitalism, industrial revolution, new economic organization and most importantly the backbone of finance; credit system. This credit system had made ways for Nepolean win, French revolution, Dutch organization, ww1, ww2, the great depression and 21'st century biggest bubble crash; financial crisis (2008 – 2009).

Commercial Capitalism

The group of merchants had been traveling all over the sea, ocean, river and desert by carrying their hardened goods to sell across for profit and sustenance. This diagram of mercantilism get exploited by commercial capitalism, where trading of goods restricted for profit generation and fluctuation of prices started happening according to their convenience's regarding their tariff's, custom, subsidy, credit, debt, costing of import and export.

[15]*Under commercial capitalism, merchants soon discovered that an increasing flow of goods from a low-price area to a high-price area tended to raise prices in the former and to lower prices in the latter. Every time a shipment of spices came into London, the price of spices there began to fall, while the arrival of buyers and ships in Malacca gave prices there an upward spurt. This trend toward equalization of price levels between two areas because of the double, and reciprocal, movement of goods and money jeopardized profits for merchants, however much it may have satisfied producers and consumers at either end. It did this by reducing the price differential between the two areas and thus reducing the margin within which the merchant could make his profit. It did not take shrewd merchants long to realize that they could maintain this price differential, and thus their profits, if they could restrict the flow of goods, so that an*

equal volume of money flowed for a reduced volume of goods. In this way, shipments were decreased, costs were reduced, but profits were maintained.

[16]Two things are notable in this mercantilist situation. In the first place, the merchant, by his restrictive practices, was, in essence, increasing his own satisfaction by reducing that of the producer at one end and of the consumer at the other end; he was able to do this because he was in the middle between them. in the second place, so long as the merchant, in his home port, was concerned with goods, he was eager that the prices of goods should be, and remain, high.

The demand and supply had created here, not technically but theoretically. Firstly, they create demand of that product by flowing goods from low to high price area and according to regional demand they restrict supply which turns prices to rise and fall and then with respect to particular value of price on demand and supply curve, point of equilibrium emerge, not just demand and curve but the idea of margin and utility also generated here, apart from transaction they exploited money and credit system also by starting lending on trade from their profit margin which returns them with compound interest, firstly from consumer side and afterwards producers side, because they play role of a middlemen, as a mediator on both they surpluses from both side by double investment and return. And secondly, with their dominance in both monetary and trading they easily spread on world map with the flag of colonialism.

Diagrammatic presentation will help readers to understand the major works of mercantilism in favor for industrial revolution, making mother countries richer and colonies slave.

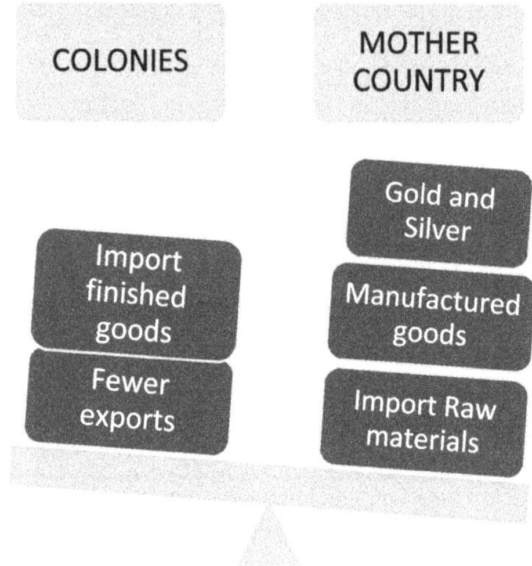

{fig, 14; explains mercantilism in brief}

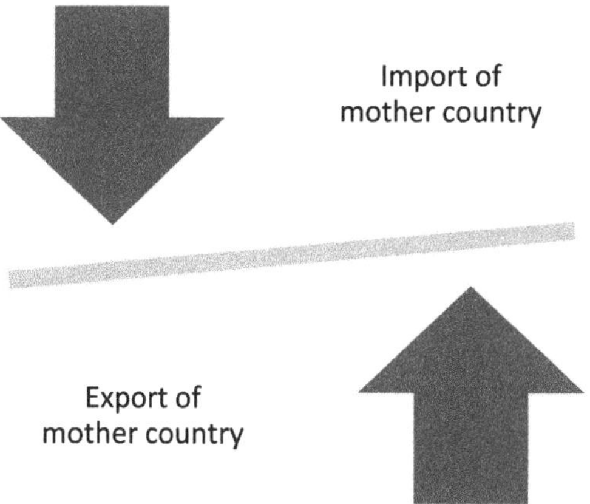

And vice versa happened for colonies, they became importing countries with only service providing in the form of labor and slavery with minimal margin of export of raw materials.

Along with these shipments of import and export, they invented one more specific idea which diagonally changed the entire financial system and build a relation between goods and money.

[17]*Some merchants began to shift their attention from the goods aspect of commercial interchange to the other, monetary, side of the exchange. They began to accumulate the profits of these transactions, and became increasingly concerned, not with the shipment and exchange of goods, but*

with the shipment and exchange of moneys. In time they became concerned with the lending of money to merchants to finance their ships and their activities, advancing money for both, at high interest rates, secured by claims on ships or goods as collateral for repayment.

[18]In this process the attitudes and interests of these new bankers became totally opposed to those of the merchants (although few of either recognized the situation). Where the merchant had been eager for high prices and was increasingly eager for low interest rates, the banker was eager for a high value of money (that is, low prices) and high interest rates. Each was concerned to maintain or to increase the value of the half of the transaction (goods for money) with which he was directly concerned, with relative neglect of the transaction itself (which was of course the concern of the producers and the consumers).

[19]The relationship between goods and money became clear, at least to bankers. This relationship, the price system, depended upon five things: the supply and the demand for goods, the supply and the demand for money, and the speed of exchange between money and goods. An increase in three of these (demand for goods, supply of money, speed of circulation) would move the prices of goods up and the value of money down. This inflation was objectionable to bankers, although desirable to producers and merchants. On the other hand, a decrease in the same three items would be deflationary and would please bankers, worry producers and merchants, and delight consumers (who obtained more goods for less money). The other factors worked in the opposite direction, so that an increase in them (supply of goods, demand for money, and slowness of circulation or exchange) would be deflationary.

The power of mercantilism had changed two most important tires of society; financial and economic tires. In financial sector, money lending has started with clear motive of credit and interest rates to enhance money in substitute of profit. People who previously trade with their produced goods and assets got dependent on bankers loan for their products shipment. Whereas in economic sector, monetary exchange triggered GDP, National income and balance of trade, the whole process of political economy get changed by this trading pattern, because of this reason many thinkers has stated that, Mercantilism was the first brick of Revolutions, either in the form of industrial, financial, economical, political or digital. But one thing was surely seen in this process of commercial capitalism or mercantilism is that each individual is concern with his only gain, need converted into greed and race of wants and desire began. As truly said by Adam Smith,

[20] *"He generally, indeed, neither intends to promote the public interest, nor knows how much he is promoting it. By directing that industry in such a manner as its produce may be of the greatest value, he intends only his own gain, and he is in this, as in many other cases, led by an invisible hand to promote an end which was no part of his intention. Nor is it always the worse for the society that it was no part of it. By pursuing his own interest he frequently promotes that of the society more effectually than when he really intends to promote it."*

NUTSHELL

Luckily, we are standing in the middle of economic street, where we not only clearly seen all previous phases of economic developments with financial thoughts but also currently ongoing scenario with the obvious showcase of future economics. And being the middle person, along with optimistic approach we have come into conclusion that the whole economy of world has been laid on the foundation of pre classical economical era only.

There are four major pickup points of pre classical economy which we seen till now on our running system with some modern nomenclature and hi-tech system. These four points are;

1. Scarce,

2. Factors of Production,

3. Allocation of Resources and Money,

4. Mercantilism/Commercial Capitalism.

1) Scarce

The one common problem which has faced by all civilization from previous phase till now is scarcity. It is a mutual discussion which had seen in early times, today's world and it will continue progressively in future also. Economists are still unable to find the solution for scarcity, for solving this puzzle many of big thinkers tried their hands but the result is still awaited.

If we examine scarcity on a broader way, it looks like a byproduct of two major components; i.e.

a) Natural components,

b) Economical components.

Natural components are the first reason of scarce, because they were limited with their limited usages creates the limited assets for many of us.

Some greater examples like, water, coal, fossil fuel, crude oil, gold, carbon and so on mining products which are limited in limited hands are fruitful for only limited persons.

Whereas Economical components, which have been largely integrated the scarce are so unlimited in the form of distribution and consumption. At this point desires get born in every human being, to consume more and more extravagantly on high supply rate, which again contributed to hike the demand rate at particular price and utility and this marginal utility creates the unlimited scarce of money.

On the first component, goods are the scarce, but on the second one, money was the actual scarce. And if both of these get combined together, it definitely forms the black hole of scarce which can't be closed or solved.

2) Factors of Production

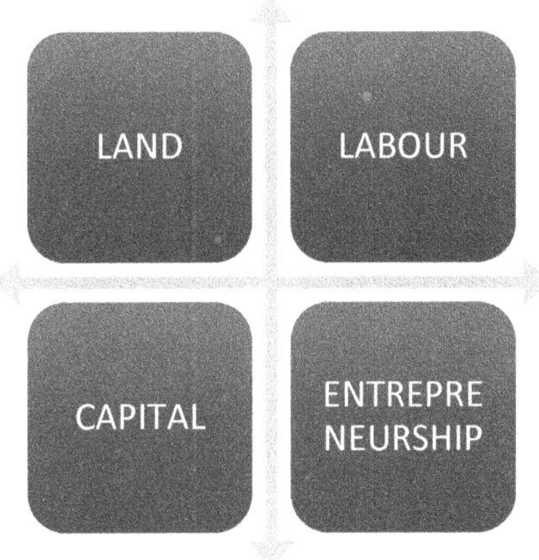

(Fig, 15; general factors of production)

There are four building blocks of economy from pre-classical times, and they all are inter-dependent on each other, by combining them on one platform we easily bring change to society.

Perhaps each of the factor also has its own consequences which again makes economy difficult to understand but the whole efforts of great minds are working uopn it which we briefly identify on further chapters where every economist of his time are seen to answer it all.

Surprisingly, this oldest model of factors of production, is now started shifting, with the replacement of some elements and enjoing of bigger elements. And believe me this upcoming new model of factor of production (which I will show you later), will definitely going to make "The Clash of Energy and Economics".

3) Allocation of Resources and Money

How to allocate resources (natural or man made), goods & services and the money are again the trouble making legacy of economic unsolved question of all the times.

Currently we have a pattern for this allocation, which quotes the circular/clock wise flow of money and the counter clock wise flow of resources in the society.

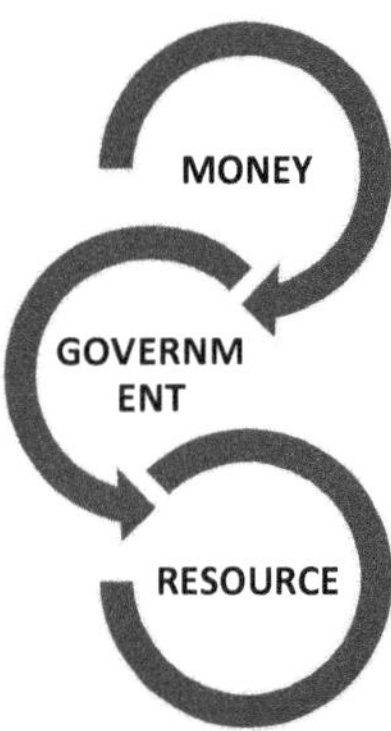

(Fig,16; Allocation of resources)

4) Mercantilism / Commercial Capitalism

The oldest continous ongoing phenomena of every particular time which is also a backbone of any society by which, none of the system nor the person/organization can run, it is the gem of Mercantilism.

Every civilization has used mercantilism for their rise, it is the general need of man to acquire some goods for him and share to others which he is not using, by this basic formula mercantilism had began and because of mercantilism only the world has became globalised in this age. Whether it is also a fact that along with mercantilistic approach commercial capitalism automatically operates in parallel direction. The trading of money than trading of goods has now became extremely powerful and we all are engaged in it very happily without even analysing what will future holds if this trading will continue.

Meanwhile these four pre-classical structures are the only contributors to further economic thought development, which can easily diagrammaticly represent as;

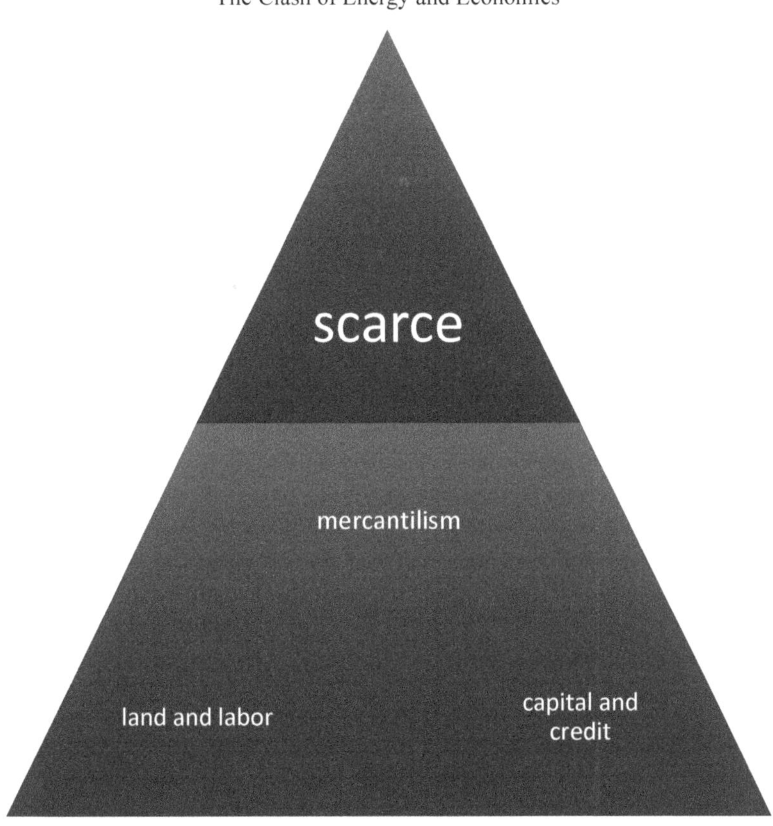
scarce
mercantilism
land and labor
capital and
credit

Classical Economy

"Necessity is the mother of Invention".

-PLATO.

Suppose you are going to start your business, so what you need first for this purpose? Of course a great business plan, where everything about your business has already predefined and define. It has a complete analysis of your balance sheet, your product, costing, target market where you are going to export, transaction and billing, rate of return and net worth, Apart from this general view there are many more things which cover your plan to successfully run your business. This is the easy game to enter for today's man, whereas that was so complicated for early man since eighteen century, but the Sun of eighteen century rises with consciousness.

The consiousness for freedom, power, victory and dominance leads this century towards inventions, at one hand inventions of industrial revolutions along with energy transformation and on the other side the necessity for economics to successfully operate the industrialization. (same like for business the need of blueprint plan). Both of these had been seen running alternately on the same age at same time to produce more, by cutting wide linear ropes of malthusian trap the classical economy or correctly say the real economy has began under the name of Adam Smith and with Adam Smith a sereies of economic expansion by verious of economists had taken place who putted their thoughts to satisfy ongoing demand and current need of industrialization to match with market and society.

I always request to those who are interested in the study of economics or wanted to start their own business, they must have to go through the journey of economics by the lens of industrial revoultion because each of the components of industry which was needed to answer were covered by economists of that time accordingly. i.e;

[21] *The classic works in the history of thought of the 1930s and 1940s then proceeded to expound and largely to celebrate a few peak figures after Smith. Ricardo systematized Smith, and dominated economics until the 1870s; then the 'marginalists', Jevons, Menger and Walras, marginally corrected Smith-Ricardo 'classical economics' by stressing the importance*

of the marginal unit as compared to whole classes of goods. Then it was on to Alfred Marshall, who sagely integrated Ricardian cost theory with the supposedly one-sided Austrian-Jevonian emphasis on demand and utility, to create modem neoclassical economics. Karl Marx could scarcely be ignored, and so he was treated in a chapter as an aberrant Ricardian. And so the historian could polish off his story by dealing with four or five Great Figures, each of whom, with the exception of Marx, contributed more building blocks toward the unbroken progress of economic science, essentially a story of ever onward and upward into the light. In the post-World War II years, Keynes of course was added to the Pantheon, providing a new culminating chapter in the progress and development of the science. Keynes, beloved student of the great Marshall, realized that the old man had left out what would later be called 'macroeconomics' in his exclusive emphasis on the micro. And so Keynes added macro, concentrating on the study and explanation of unemployment, a phenomenon which everyone before Keynes had unaccountably left out of the economic picture, or had conveniently swept under the rug by blithely 'assuming full employment'.

If we filter the phase of classical economy, we got two majorly huge names which apparently never sit together, in fact they can't be named together, by putting them together is like putting two swords in one sheath, because they both have completely different ideology and based on their ideology they both contributed in the making of two super powers along with two giant system's. They are;

> A. Adam Smith,
>
> B. Karl Marx.

A. <u>Smithian's Economy:</u>

Its 1776, the year of two big events; America got the success on its struggle and declared the independence. Meanwhile a treatise named "The wealth of Nations" by Adam Smith get published and made him the father of modern economics which touches and torches modern concepts, providing pillar of Capitalism from where privatization starts.

Smith's theory gives us three main areas or factors of economics;

> **a)** Division of Labor, which is also has sub-unit, called; wages and rent,
>
> **b)** Division of Stocks/capital again has sub-unit, called; Profit and interest rates,

c) Lastly and more importantly the concept which he continued from pre-classical era, The Mercantile system, which is Centre of his thought and theory. He called it Commerce, and it's again having sub-unit called; price and market.

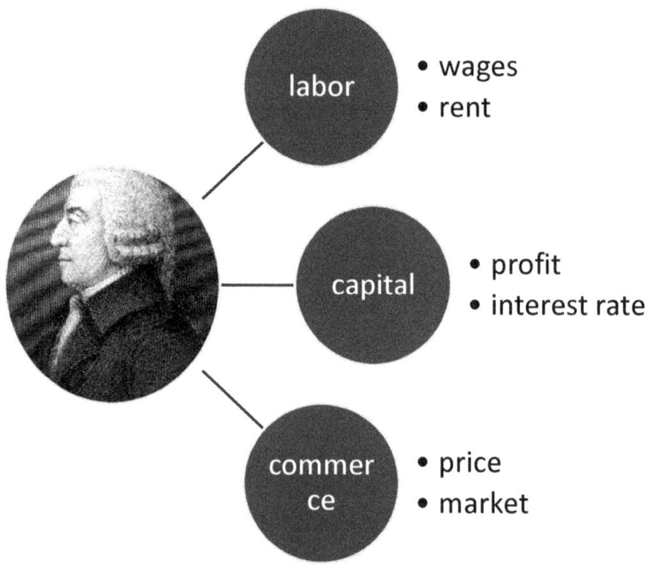

(Fig, 17; explaining smithian's thought.)

B.) <u>Division of labo</u>

He starated his view with the correct imployment of labor. Any nation having any of its circumstances can begin to grow with one brilliant decision that is, correct employment of labor according to their skills, production of any country is so dependent upon the division and accumulation of skilled labor.

On this frame we make our formuale here;

Say, P- is the production of goods (any country) and L- is labor for that production; therefore we have;

$$P \propto L \quad \ldots\ldots\ldots\ldots \text{①}$$

Production of goods is directly proportional to its labor.

[22] The Philosopher giving an example of pin-making industry, he explains, how the educated and team working power of labor making four thousand eight hundred pins in a day per person. The productive power of labor can achieve with the proper division and combination of labor.

But this labor theory demand one thing which was the biggest milestone for every economists, and also it is a common factor which has discussed by all of them, it's wages.

a) Wages;

Every human wants a payback of his efforts, it is a proven fact that wages are the only thing by which labor gets attracted, it gives them incentives towards production and also by this competency created among them for growth. Although it is fluctuating, according to specific time and market regulatory forces but still works for production of labor.

[23] Though the wealth of a country should be very great, yet if it has been long, stationary, we must not expect to find the wages of labours very high in it………

We didn't find him in the favour of high wages, because in his view, increase in wages, increase the price of commodity.

It concludes that;

L α W …………②

Denots labour is directly proportional to wages as our first equation,

By equation 1 and 2;

P α L ……………①

L α W ……….…②

We have;

$$P \quad \alpha \quad W$$

It means that production of any nation is not directly proportional to labor but to wages. If this factor experience any decline it directly leads to the fall of production and unemployment. Just exactly had happened in the USSR, the biggest collapse of Communism is also because of wages, labour had already aware about his incentives either he work hard or not he will get his fixed income, no increments, no bonus, not a single share in property or asset, made labours lazy and non productive towards society.in fact they became first audience againt Communism.

b) Rent;

After wages, another component which doubles the cost of production is rent by landlords. Rent with the combination of wages and profit decide the market price of nation.

Price (Pr) = W + R + Pt.

(where R is rent and Pt is profit)

This equation plays critical role in any economy because they each inter-dependent on each other, slight change in any of us make whole equation change, increase in any factor makes increase in all and same decrease in any one automatically decline all.

We summarize it like;

Pr = W + R + Pt

$\uparrow$Pr = $\uparrow$W +$\uparrow$R +$\uparrow$Pt ......③

$\downarrow$Pr = $\downarrow$W+$\downarrow$R +$\downarrow$Pt ....④

c)   Division of stocks/capital ;

[24] *The general stock of any country or society is the same with that of all its inhabitants or members and therefore naturally divides itself into the same three portion, each of which has a distinct function of office.*

1.   *First, portion reserved for immediate consumption, it affords no revenue or profit.*

Eg; food, clothes, household furniture etc, which have been purchased by their proper consumers.

2.   *Second, is the fixed capital of which the characterstic is that it affords a revenue or profit without circulating or changing masters.*

Eg; land, machines, warehouse, buildings and talents.

These are the assets, which has possession or ownership.

3.Thirdly, circulating capital, which affords a revenue only by circulating.

Eg; money, finished goods etc.

It means,

C = X + Y + Z ....①

C _ capital

X _ reserves

Y _ fixed asstes

Z _ circulating stocks.

Profit is generating only by Z and some of Y, that is why all the efforts of economy are made for this factor (Z) , because Nation's GDP are revolving around this macroeconomic factor, this game is only played by huge investors and market regulators. whereas in Y only some of the people who have ownership in specific assets which are tangible and producing all the goods through their medium of assets having key role here again they are majorly industrialists, only they are allowed to take advantages of this revenue. Lastly and silently remain X who is only a consumer on a micro level have no hand in productions, so no share in revenue.

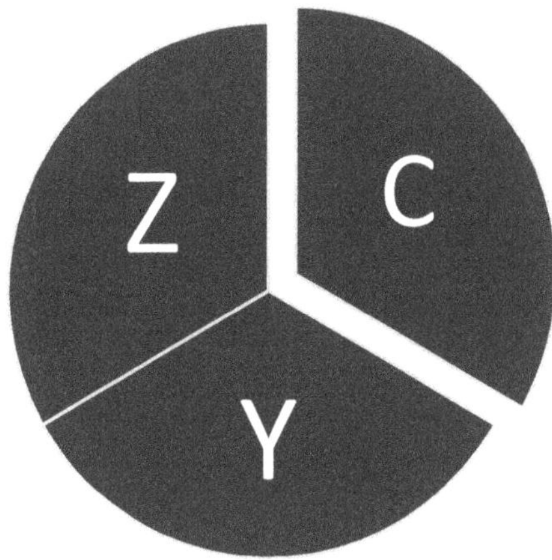

(Fig, 18; capital accumulation)

By analysing the pie chart of capital its clearly seen that in our running system capital is very restricted with the restriction of accumulation. As we saw in accumulation of labor, correct employment and team work is the thing which helps in production at higher rate. This same pattern should also be need to applied in accumulation of stock.

Can we afford to sit on just one labor for production, then how we are sitting here when profit is concern.

a) **Profit;**

The most attractive concept of not just thinkers but all human being. We all want profit in our way, any how. This any how, in economy gets more trickier. The profit of stock very with the price of the commodity in which it is employed. This employment of stock are already limited as we saw the running capital is only Z. hence, for profit maximisation on this limited resources, we obviously need two things; which are also considered as major resources of revenue;

I. Price,

II. Interst Rates.

[25] *According, therefore, as the usual market rate of interst varies in any country, we may be assured that the ordinary profits of stock must vary with it, must sinks as it sinks, and rises as it rises. The progress of interest, therefore, may lead us to form some notion of the profit.*

So, we have the formulae of profit maximization;

$Z \rightarrow IR$ (IR- interest rates)

And for fixed asstes we have interst in compound form;

$Y \rightarrow IRc$ (IRc – compound interst rates).

Therefore, The rate of interest is the actual face value of stocks/capital for revenue. Hence, equation 1) will get change here, we have renew formulae of capital;

$C = X + IR + IRc$

b) Interest Rates;

Though we had a long history of interest rates, unfolding debt and credit generally seen in every system, but we have been never find once in a history of interest rates, that any system, community or organization never get failed by using formulas of interest rates, In fact those who have billions of net revenue, Also carrying billions of debt on market cap.

Those capital who lent on interest provide quicker results by doubling and tripling of capital but again there is some kind of magic in interest rates, which reversably diminish the output of that particular stock. As the philosopher also briefly states that;

[26] *In proportion as that share of the annual produce which as soon as it comes either from the ground or from the hands of the productive labourers, is destined for replacing a capital, increase in any country, what is called the monied interest naturally increases with it. The increase of those particular capitals from which the owners wish to derive a revenue, without being at the trouble of employing themselves, naturally accompanies the general increase of capitals; or, in the other words, as*

stocks increases, the quantity of stock to be lent on interest grows gradually greater and greater.

As the quantity of stock to be lent at interest increases, the interest, or the price which must be paid for the use of that stock necessarily diminishes, not only from those general causes which make the market price of things commonly diminish as their quantity increases, but from other causes which are peculiar to this paarticular case. As capital increases in any country, the profits which can be made by employing them necessarily diminish.

It becomes gradually more and more difficult to find within the country a profitable method of employing any new capital. These arises in consequence, a competition between different capitals, the owners of one endeavoring to get possession of that employment which is occupied by another…. their competition causes the wages of labourers and sinks the profits of stock. But then the profits which can be made by the use of a capital are in this manner diminished, as it were. At both ends, the price which can be paid for the use of it, that is, the rate of interest, must necessarily be diminished with them.

The best example of Interest rates are like of, that tree which gradullay starts growing and growing but this tree is growing without fixing in soil, this tree doesn't laid on surface, it just flows in air and whenever market have some kind of greater forces towards this tree on interest rates, it gets down and diminished, like it had been never existed before.

So, this is the diminishing story of interest rates, which provides result, highlighting the fact, why start-ups mostly get failed, who entered with credit based business system, they get diminish in same way as interest rates. This science will be covered with the correct business plan in last sections of that book.

d) Commerce/Mercentile;

Undouble, he is the promoter of mercantile system with the commercial capitalistic approach. A free trade is major part of his thesis, more over his central focus vectors on Macroeconomics.

However, market and price are key ingredient of mercantile society.

The influence of mercantalist in Smith thought has it's peak, that in easily reflect in his given definition of economics; by defining economics also he loudly highlighted merchants and their role.

This definition of economics also widely used in modern day economy, it states that:

[27] *It cannot be very difficult to determine who have been the contriver's of this whole mercantile system, nor the consumers, we may believe, whole interest has been entirely neglected; but the producers, whose interest has been so carefully attended; and among this latter class our merchants and manufacturers have been by far the principal architects.*

I have made a diagram, which completely explains Smithian's commerce by analyzing his commentary.

(Fig, 19; shows upward moments of commerce)

As the commodity from raw sends to upward level for more produced form, there is a rise and on each rise again generated rise in prices, due to this, he said and used the term invisible hand, which is nothing but a market and its forces. The architect of this whole economy, sets the market price by leaving nominal price, which fixes level of supply in cordination with level of demand.

a) Market;

Markets have centrifugal force in the society, any nation starts with market and its invisible forces which flows the free economy although in my opinion its clearly visible if apply litmus test here, and litmus of market is 'Quantity'.

Quantity of every particular commodity, regulates market price, which is far from actual natural price of commodity. Market price is the addition of natural price, quantity of commodity and quantity demad of that commodity which all bought to market.

So we make formula here;

Market Price(MP) x' = Q + Qd + x.

[28]*When the quantity bought falls short of the effectual demand, the merket price rises above the natural price.*

Therefore;

X' > x

When the quantity bought exceeds the effectual demand the market price falls below the natural price.

Therefore;

X' < x

b) Price;

As we previously obtained formula of price;

Pr = W + R + Pt

When wages increases with the increase of stocks, it increases the liabilityies of entrepreneur also with the increase in price. His profit ratio gets cut down, half in paying the cost and expenses, some in clearing the deferred payments, because interest comes with liability and when in this case wages increase automatically lowers the profit ratio.

For defining the actual price of commodity, we have to add all the components which had been served for commodity, we already added cost of wages and rent but how and why profit is added on price system?

[29]' *The real value of all the different component parts of the price, is must be observed, is measured by the quantity of labour which they can each of them purchase or command. Labour measures the value not only of that price which resolves itself into labour, but of that which resolves itself into rent and of that which resolves itself into profit.*

Businesses in current market system runs on borrowing, where borrower has to repay his lender after particular time being, this further repayment is simply called; interest rates, which has to be included in price mechanism. Because of its uniqueness in business, entrepreneurscant neglect that in their pricing specially if they are venturing on credit. Why?

Smith replies;

[30]*The interest of money is always a derivative revenue, which if it is not paid from the profit which is made by the use of the money, must be paid from some other source of revenue, unless perhaps the borrower is a spend.., who contracts a second debt in order to pay the first.*

Concludes that, profit should always be divide by interest first, to get the clear accountability either for revenue or for price. But the saddened part of this methamatics, lowers the revenue and raises the price.

Hence our modified formula of price, which is majorly applied on credit market, now looks like;

Pr = W + R + Pt/IR

This is the reason why we found Smith in favour of surplus, for organisational and national growth.

[31]*The lowest ordinary rate of profit must always be something more than what is sufficient to compensate the occasional losses to which every employment of stock is exposed. It is this Surplus only which is neat or clean profit. What is called gross profit comprehends frequently, not only this surplus, but what is retained for compensating such extraordinary losses. The interest which the borrower can afford to pay is in proportion to the clear profit only.*

Surplus always has a positive role in economy and in investment. But how we are going to obtain it? Let's ask one more thinker of that same time.

Marxists' Economy

On the surface of Classical economy, one more name had emerged outstandingly, although this man had been collected many critics and hatered because of his anti-capitalistic approach but then also his menifesto had made its position that evry student or learner of economy has to go through with his menifesto, even though the great economists of west also, reading and applying some of his factors by filtering in the system.

This huge manifesto compiled in three volumes, somestimes called communist manifesto with the title of "Das Capita" by Karl Marx.

After Adam Smith, Marx also had achieved its influence on economy. If we analyse his fat volumes, it will give crystall clear reflection of previous factors, as discussed by Smith. Not only in Marx's commentary but in all economists work, we will find same subjects repeatedly discussed, only difference of structures they had made for individual chapters.

The complete thesis of Marx, also consists of

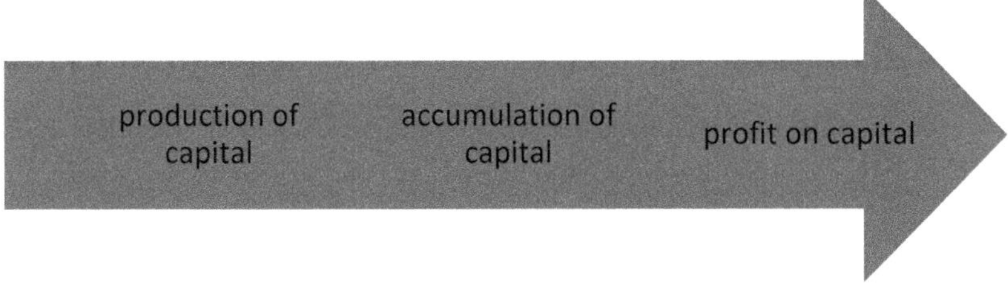

(Fig, 20; economic pattern of Marx)

Along with the symbol of hammer and sickle, marx had came upfront in favour of producers rather than merchants, by promoting labours and producers of first stage, he presented a circuit;

Where, circuit 1: works in favor of producers. Perhaps, circuit 2: goes for merchants.

Apart from circuits, he also argues on market system of profit maximization and its crises. His major focus was microeconomic factors, which now get converted and use in macro level by modern philosophers.

- Circuit 1; C-M-C

[32]*The simplest form of circulation of commodities is C-M-C, the transformation of commodities into money, and the change of money back again into commodities; or selling in order to buy.*

This applies for producers, in where they can have some sorts of income by selling their produced commodities, to buy their wanted goods. But this pattern will not going to generate great profit.

- Circuit 2;

M-C-M

[33]*The transformation of money into commodityies, and the change of commodities back again into money; M-C-M, or buying in order to sell. Money that circulates in the latter manner is thereby transformed into becomes capital, and is already potentially capital.*

This circuit was always lovable to traders, its also called as money dealing capital, hugely generate profits, but here again same problem arises as arised in smithians thought. This profit has to gone through the process of long deductions, thanks to interest rates.

He prolonged money dealing or, M-C-M, because in this system, exploitation of money will occur, which is counter part of his thought.

producers again rest with empty hands by this mechanism, without having any percent of share in profit through their commodity.

[34]*Money dealing in its pure form, which we consider here, i.e, set apart from the credit system, is thus concerned only with the technique of a certain phase of commodity – circulation, namely, that of money circulation and the different function of money arising in its circulation.*

This substantially distinguishes dealing in money from the dealing in commodieties, which promotes the metamorphosis of commodities and their exhange or even gives this process of the commodity – capital, the appearance of a process of a capital set apart from industrial capital. While, therefore, commercial capital has its own form of circulation, M-C-M, in which the commodity changes hands twice and thus provides a reflux of money, as distinct form C-M-C in which money changes hands twice and thus promotes commodity exchange, there is no such special form in the case of money dealing capital.

It is equally evident that the money dealers profit is nothing but a deduction from the surplus value, since they operate with already realised values (even when realised in the form of creditors claims).

Market again regulates on credit system, by M-C-M model; which states that, businesses again get started with borrowed money, in the form of credit, for particular time. With these borrowed money, trader or merchant perform some kind of trading by supplying demanded goods/commodities to consumers/buyers. And he tried to cover all his costing from consumers, by adding all factors of profit, at the end of transaction, he will be able to achieve his profit, but at this point comes the trickiest part of credit system, your time limit of borrowed money got over and you have to repay the principal amount along with the multiple charges of time, till you had the principal amount. Profit, cuts down and you only left with some money which will never be sufficient for further investment and trading, so again this cycle repeat.

If any entrepreneur, who is willing to start his venture without fall, along with profit and surplus values, zero payments to lender and free market operations by the addition of securitizations, they need to understand, three sciences;

     I.   The Science of Interest rates

     II.   The Science of Business and Credit Cycle

    III.   The Science of Time.

They all get covered, with clearnace in our further chapters of BOOK1. Meanwhile we collect general ideas of economic giants, how they applied these sciences in their time, and how much benefit they get from these economic science. so that we make error-free module for coming entrepreneurs, which will discuss on my BOOK2 with renewed economic thought and The Blueprint, specially for generating SURPLUS.

❖ Marx on Interest Rates

[35]*To determine the average rate of interest we must; 1) calculate the average rate of interest during its variations in the major industrial cycles and 2) find the rate of interest for investments which require long term loans of capital.*

Lets say x' is the rate of interest for 10 years,

And, average rate of interest;

x1+ x2+ x3 +x4….x10/10

by this long term calculation, only one thing reaches out at height, which is debts.

[36]*The rate of interest reaches its peak during crises, when money is borrowed at any cost to meet payments. Since a rise in interest implies a fall in the price of securities, this simultaneously offers a fine opportunity to people with available money – capital to aquire at ridiculously low prices such interest bearing securities as must, in the course of things, at least regain their average price as soon as the rate of interest falls again.*

Venture of investments get multiplied by compound interests, which leaves entrepreneurs out of revenue (by means of surplus). They just got some shares of income at the end of balance sheet.

Conclusion: Classical Thoughts had failed to answer Surplus Theory. We now moved to Neo-Classical thought for the searching of surplus value.

First Wave of Energy and Economics

Since 1800, energy and economics, both has started working together, for each other and for combined purpose and cause, which is to create a continous flow of industrial, developmental and advanced waves for every changing world. In its first wave, which is also a fundational wave of unsustainable energy, where not only transformations of energy and economics began but also developmental era and colonialsim started, and they all started happening together, because of one same reason, which is the chain reaction of energy and economics.

During first wave, obviously, everything is new, fresh and required lots of hardwork in everyforms of aspects, whether; its labour work, capital, organization, accountability, land, managements of everything, everything needs special attention with careful handling, because it was the first experiment so its needed to be successful, for furthermore. Which exactly happens with anyone of us, who has intention of start-ups, put yourself in that position and answer, what are the thing you need to do for your start-ups.

Here, start-up of steam, engines, railway's and cotton had occurring simultaneously. This start up had been revenued huge output, along with the formation for second wave, it has promoted industries and organization in later century also.

This first wave had given tremendous results, because of three factors;

- Product selection

- Mode of transaction

- Principle.

Product selection:

Early phase of revolution targetted most effective product, which is cotton. In these time, people or craftsman; using their skills of handcrafts in effective manner to produce cotton first, and then transform in into garments by passing it from dying, drying and designing. Garments was the primary need of people after food, as population increase, demand of garments also increased accordingly, perhaps early method had been failed to produce more and satisfy the demand, meanwhile this techniques had been operated by only sustainable form of energy at that time but, whenever needs and wants of human comes, it ignores everything for their selves.

Spinning wheels turned into machines, this turning point of hand process to machine process, increases the demand with satisfactory supply of cotton and garments. Cotton had played most significant role for boosting first wave, but; here cotton is not only product which contributed for first wave, invention of first spinning machines for the production of cotton, was the actual product, with wise invention and productive transformation of primary renewable resources of energy, called; hydropower. By the help of hydro-power plant, first cotton spinning machine, produced unexpected turnover, along with steam engine.

(fig, 21; explaining and showcasing, first water- powered spinning mill, which is only build by primary resources of energy.)

Mode of Transaction;

After product in any businesses, comes the second obstacle which is transportation. This part of business, required huge amount of investment in transporting goods, with securities, tariff's, international taxes, costing of fuels, labours, freight charges and services. Although, its all added to consumer's invoice, but meanwhile buyer and supplier both get tied up with some financial institution, and regulate according to them for ease of payment and security.

Indeed, some extraordinary invention had been taken place on the first wave which has never been seen before, it's the horn of railways, by the use of steam engine. Here again, primary resource applied for generating, transformative energy. i.e; heat and water, which afterwards get combined with some chemical elements like coal, to act as fuel and power supplier for railways and also helpful for generating second phase of industrial revolution.

Because of railroads, transactions happened easily on less time. And this kind of tranquil transactions, began the era of economic globalization.

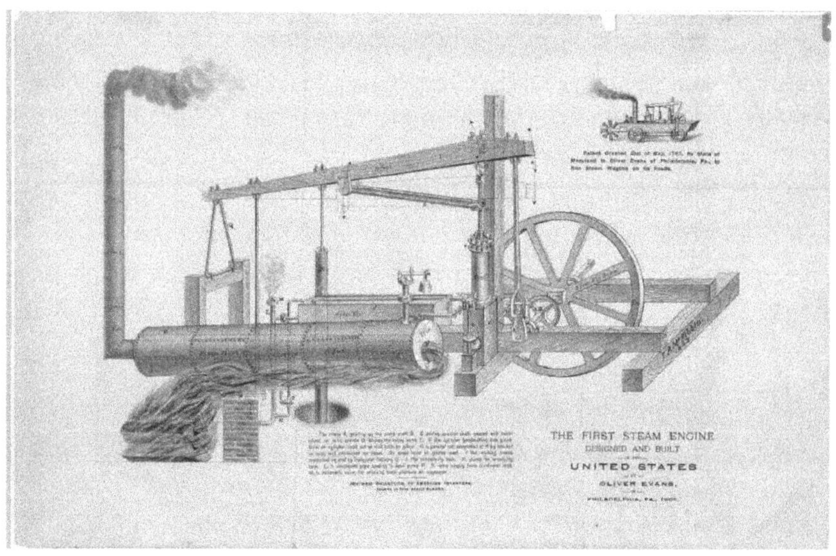

(fig, 22; represents the diagram of first steam engine.)

Principle

Mathematics, will takes place for every working thing, here also in the first wave of energy and economics, we observe a very define and calculative mathematics in every aspects, which principle the whole system and this principle has become manuscript for later waves.

The first principle, applied here was the combination of producer and worker; where, the production of goods in industries by the help of renewable energy get combined with some economic factors of production, namely; labor, land, wages, rent, capital, organization and revenue. Most of the thinkers of early and classical times of economics were involved, engaged in dragging and sorting these concepts only.

Smith, Ricardo, Mill, Marx, and many others had mostly argued on first principle, these first principles, tremendously matched with first wave of energy and economics, like DNA, and because of this double helix DNA structure, eighteen century acquired successful result of first industrial revolution.

After the success of first principle in economy, which helped society and nation to grow higher at per capita rate, many principles, mathematics and calculations, had started discussing among economists, many new principles get added in the field of economics, and then after the end of eighteen century, we witness; "Principles of Economics."

Neo – Classical Thought

Neo-classical economic thought had been came into existence, during the late nineteenth century, in late 19's several important changes took place in world's economy. The second industrial revolution caused growth in industry and transportation, which allowed increase in trade between nations globally. European industry expands and Germany became the new leader. The new born old United States also enjoyed success during the second phase of industrial revolution. In fact, industrialization had made one year old U.S, the richest nation in the world at that time.

Today's global economy is simply a description of the integration of trade in goods, services and money world-wide. The spread of trade and investments in abroad is linked with a process called, Imperialism occurred between 1870 and 1900, which also labeled as European Imperialism.

Imperialism demands huge statistics, data analysis and wide range of margins and these riddles exceptionally solved by father of modern microeconomics, Alfred Marshall.

Marshall is the man who took economics to more advanced and mathematically rigorous level. He defined, *"Economics is the study of Men."* Under this definition, Behavioral Economics had born, where analysis of market had performed with the analysis of individual's behavior. On certain circumstances with difference of conditions and time, how individuals would be going to behave, moreover; along the pattern of behavior economics, assumption based calculations, math, statistics and market prediction starts appearing in neo-classical age.

Alfred Marshall

Marshall was the first economist who developed, the standard supply and demand graph which represented his whole economic thought towards society, market and individuals.

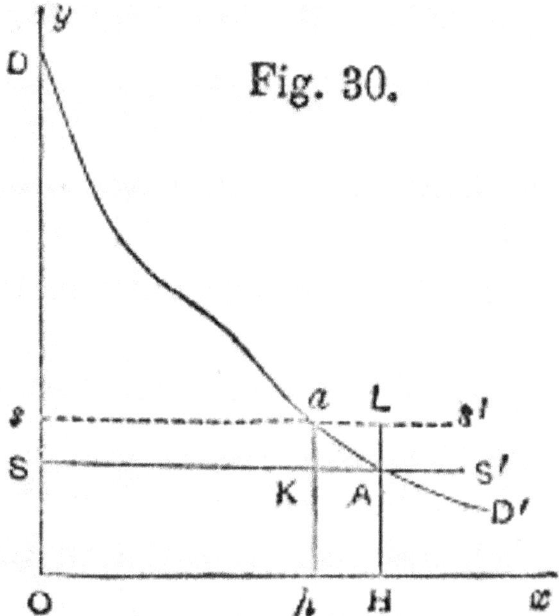

(Fig, 23; Marshal's fundamentals on principle of economics)

This Graph represents number of fundamentals:

1. Supply and demand curve,

2. Market equilibrium _ceteris paribus technique,

3. The relationship between quantity and price in regards to supply and demand,

4. Law of marginal utility, diminishing returns and interest on capital.

5. Idea of consumer and producer surpluses,

6. Credit and cycle,

7. Time factor.

It all starts with demand and quantity demand, according to demand of market, monopoly of supply generated. Meanwhile price act as bridge in between the quantity demand and supply.

[37]*When a trader or a manufacturer buys anything to be used in production, or be sold again, his demand is based on his anticipations of the profits which he can derive from it. These profits depend at any time on speculative risks and on other causes, which will need to be considered later on. But in the long run the price which a trader or manufacturer can*

afford to pay for a thing depends on the prices which consumers will pay for it, or, or for the things made by aid of it. The ultimate regulator of all demand is therefore consumer's demand.

[38]*His demand becomes efficient, only when the price which he is willing to offer reaches that at which others are willing to sell.*

[39]*The prima facie interest of the owners of a monopoly is clear to adjust the supply to the demand, not in such a way that the price at which he can sell his commodity shall just cover expenses of production, but in such a way as to afford him the greatest possible net revenue.*

There are so many variables involved in progressed economy, which are simultaneously working together at a time and because of each variable, reflux of others unwantedly achieved.

Marshall suggested constant variables for improving involuntary outcomes, and working on one thing at specific time. This can achieve by limiting one particular problem on equilibrium framework by keeping most of the variables constant for that specific time period and then slowly permitting other things to vary, but for performing this we have to carefully use time method, along with marginal utilities, diminishing returns based on interest and credit cycle.

[40]*It is true that in almost every business there is a constant increase in the amount of capital required to make a fair start: but there is a much more rapid increase in the amount of capital which is owned by people who do not want to use it themselves, and are so eager to lend it out that they will accept a constantly lower and lower rate of interest for it. Much of this capital posses into the hands of bankers who promptly lend it to anyone of whose business ability and honestly they are convinced. To say nothing of the credit that can be got in many businesses from those who supply the requisite raw material or stock in trade.*

[41]*When a man is engaged in business, his profits for the year are the excess of his receipts from his business during the year over his outlay for his business. The difference between the value of his stock of plant, material etc. at the end and at the beginning of the year is taken part of his receipts or as part of his outlay according as there has been an increase or decrease of value, what remains of his profits after deducting interest on his capital at the current rate (allowing, where necessary, for insurance) is generally called his earnings of undertaking or management. The ratio in which his profits for the year stand to his capital is spoken of as his rate of profits. But this phrase, like the corresponding phrase with regard to interest, assumes that the money value of the things which constitute his*

capital has been estimated: and such an estimate is often found to involve great difficulties.

Surplus is again unsolved question here, either for consumers or suppliers. They both failed to generate surplus value according to their asset value, by entering on the field of business; they had entered into maze of money, where monetary system, interest based borrowings, credit default swaps, inflation, deflation, market fluctuations, they all get united against surplus value of the society and shrink it to limited hands only.

Whether, normal people just engaged themselves to generate profit, taking their profit out from running money market by deducting all necessary expenses including market and interest cap, is primary goal of common people. When their business is begging for profit only, then how come they will have stocks of surplus.

[42]*When we come to discuss the money market we shall have to study the causes which render the supply of capital for immediate use much larger at some times than at others; and which at certain times make bankers and others contented with an extremely low rate of interest, provided the security be good and they can get their money back into their own hands quickly in case of need. At such times they are willing to lend for short periods even to borrowers, whose security is not of the first order, at a rate of interest that is not very high. For their risks of loss are much reduced by their power of refusing to renew the loan. If they notice any indication of weakness on the part of borrower; and since short loans on good security are fetching only a normal price, nearly the whole of what interest they get from him is trouble. But on other hand such loans are not really very cheap to the borrower: they surround him by risks, to avoid which he would often be willing to pay a much higher rate of interest. For if any misfortune should injure his credit, or if a disturbance of the money market should cause a temporary scarcity of loanable capital, he may be quickly brought into great straits. Loans to traders at nominally low rates of interest, if for short periods only, do not therefore really form exceptions to the general rule just discussed.*

Not just money market theory, if look forward for their causes we will find mutual factor in every case, which is;

[43]*When we come to discuss the causes of alternating periods of inflation and depression of commercial activity, we shall find that they are intimately connected with those variations in the real rate of interest which are caused by changes in the purchasing power of money. For when prices are likely to rise, people rush to borrow money and buy goods, and thus help prices to rise; business is inflated, and is managed recklessly and*

wastefully; those working on borrowed capital pay back less real value than they borrowed, and enrich themselves at the expense of the community. When afterwards credit is shaken and prices begin to fall, everyone wants to get rid of commodities and get hold of money which is rapidly rising in value; this makes prices fall all the faster, and the further fall makes credit shrink even more, and thus for a long time prices fall because prices have fallen.

Two common factors, namely; credit cycle and interest rates, are always seen in each cause: either it is inflation, deflation, depression, financial crises, or, revenue generation, low surplus and price fall.

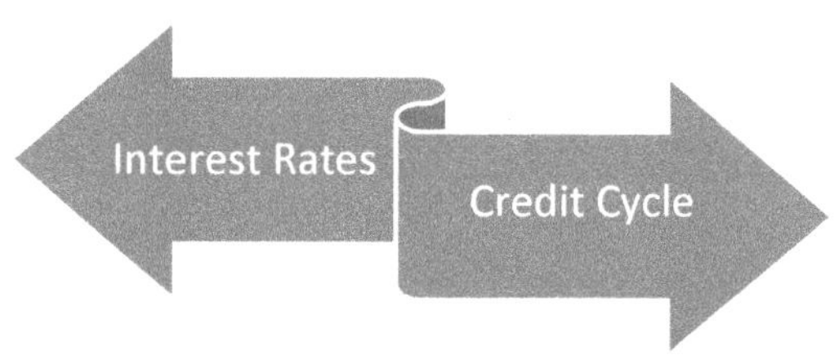

(Fig, 24; regulators of price)

They both work separately and alternatively to produce more kinds of results to some elite class; and more kinds of tricky troubles to all people. Marshall, in his commentary pointed out one specific vector in economy, which always and always has the key role for balancing everything because of its elastic nature on inelastic market and that is unquantified Time.

[44]*The various time periods are defined in terms of the economics of the firm and of supply. The market period is so short that supply is fixed, or perfectly inelastic. There is no reflex action of price on quantity supplied, as the period is too short for firms to be able to respond to price changes. The short run is a period in which the firm can change production and supply but cannot change plant size. Here there is a reflex action, as higher prices cause larger quantities to be supplied and the supply curve slopes upward. In the short run, the total costs of the firm can be divided into two components: costs that vary with output, which Marshall termed special, direct, or prime costs and modern texts call variable costs; and costs that do not vary with output, which Marshall termed supplementary costs and modern texts often call fixed costs. The distinction between variable and fixed costs in the short run was evidently drawn from Marshall's observation of the business world. It became an important*

analytical tool in analyzing the actions of the firm. In the long run, plant size can vary and all costs become variable. The supply curve becomes more elastic in the long run than in the short run, as firms are able to make full adjustment to changing prices by altering plant size. The long-run supply curve for an industry can take three general forms: it can slope up and to the right (costs may increase); it can be perfectly elastic (costs may be constant); or, in unusual situations, it can slope down and to the right (costs may decrease). The secular period, or very long run, permits technology and population to vary, so Marshall used this construct when he analyzed the movement of prices from one generation to another.

Conclusion: In Neo-Classical thought too, answer of surplus still unsolved. We lastly search it on the modern economic thought.

And after the discussion of modern economics, we will reframe the economy of global time in BOOK2, where we have all basic ingredients of economics more specifically, SURPLUS.

Second Wave of Energy and Economics

Electronics and electricity is the byproduct of second wave of energy and economics. Second wave is also considered as most influential period for modern world, during this period industrial revolution exceeded from the boundaries of western world and speedily spread in eastern demographics with the huge inventions of telephones, televisions, bulbs, radio's and here we welcomed digital world.

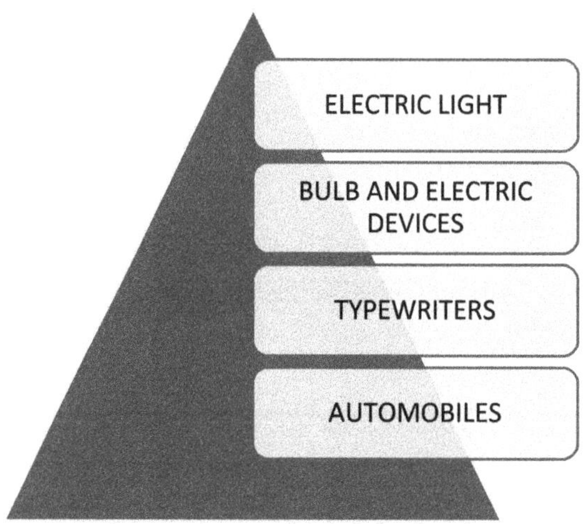

(Fig, 25; inventions of second waves)

When energy of nonrenewable resources, coal, fuel and hydropower plants get collide with the economies of margins, utilities, quantities demand and supply, inflation rate and credit, interest cycle of time, it had produced huge amounts of;

1. Mass Production,

2. Consumerism,

3. Need of a medium.

Mass Production

Inventions lead the society towards production, but; inventions of energy along with the inventions of money trigger mass production. Factories are getting well established already, on the phase of second wave, it transforming itself towards more automation.

Labor – wages theory influenced manpower in factories which automatically promoted production, in addition of manpower, organizational approach of industries also stepped up to growth, with the continuous supply of capital in industries either in the form of equity or debt, in both cases it helped industries to revolve around. Rate of production raises higher and higher with the expansions of quantity supply in the market, which again make market rate high to increase prices with the increase in gross domestic product (GDP) to increase in inefficiency of economy amidst chaos of people also increases which had given birth to behavioral economy to understand.

Consumerism

When rate of production increases spontaneously and simultaneously it increases with the increase in demand of consumers. Want takes place the position of need and willingness to purchase gets higher. These two things are responsible for making our quantity demand supply curve to more curves and inefficiency which doesn't allow equilibrium state to market. Whenever equilibrium is going to achieve at that point only, this curve get stretch towards either to upward or downward direction.

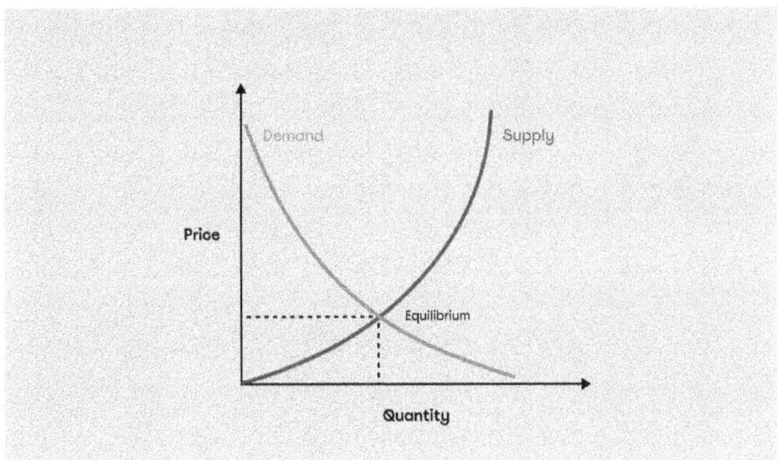

(This figure, 26; represents Basic quantity demand and supply curve)

A marginal utility in economics becomes the center point to study because of consumerism only.

Both consumerism and mass production required one kind of medium to exchange each other and enjoy their specific benefits, this medium has now covered not only economy but the whole world in them and made man an economic animal.

Need of Medium

The most fascinating part of history has found in the second wave of energy and economics which is also the major key reason of biggest empires; this is the era of Gold Standards.

For understanding gold standards, firstly we have to understand the history of monetary system; which medium of monetary system had been used along all the times and why they get change with respect to time.

[45]*Historically, the silver standard and bimetallism have been more common than the gold standard. The shift to an international monetary system based on a gold standard reflected accident, network externalities, and path dependence. Great Britain accidentally adopted a de facto gold standard in 1717 when Sir Isaac Newton, then-master of the Royal Mint, set the exchange rate of silver to gold too low, thus causing silver coins to go out of circulation. As Great Britain became the world's leading financial and commercial power in the 19th century, other states increasingly adopted Britain's monetary system.*

A fixed quantity of gold, has measured as economic unit and medium of exchange to perform economic activities also termed as monetary system. In this monetary system gold has used as medium that's became the reason of its name as gold standards. Before gold standards or before 1873, it was silver for millennia on which all the economic activities had performed, all prices are set on silver rate and exchange this currency of silver.

Here, one question arises, that what makes need of this medium to occur when there is already one monetary system is existing in the society. Why people rushed for gold when they already had silver, it might happen because of three reasons;

First; value of currency increased by gold in comparison with silver, second; it also act as stockpile to governments, and lastly; imbalances of trade also get settled by gold all over the globe. Hence, England the founder of industrial revolution again becomes the first one to adopt gold standards to make his position clear on world.

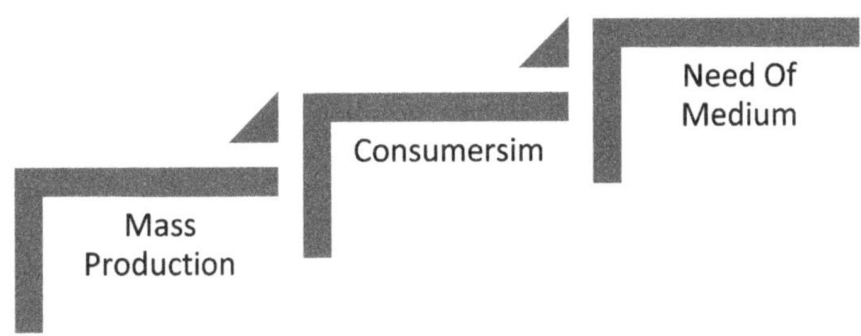

(fig, 27; explaining influential second wave of energy and economics)

Modern Economic Thought

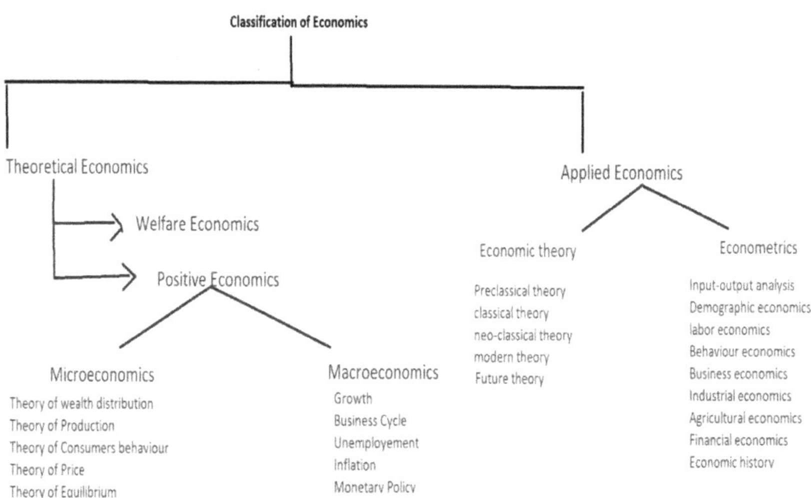

(Fig, 28; modern classification of economics)

It's the beginning of new world order with the new dynamics of operational methods controlling worldly orders in each sectors of life. Our mother earth from the new decade of 20th century has rapidly changed in various of terms specially in the term of population, the unprecedented rate of population growth, in the beginning of this century had started with 1.6 billion people, and ended with around 6.2 billion people. Surely, this hike of population demands hike in finances for maintaining human index rate and for doing this all we all stepped in to modern economic thought in widely elaborated form in where theories and practice both added together to produce more kinds of results.

Major events of modern economic world include decolonization, nationalism, globalization, new forms of intergovernmental organizations, international corporations, information and communication technology, with the spread of democracy, women voting rights, influence of western culture, and continuous chain of wars till the end of 20th century.

John Maynard Keynes;

World War 1, 2, nuclear power, cold wars, post- cold war, digital revolution, the great depression, collapse of two giant super powers of that time; USSR (Soviet Union) and Ottomans. These series of events taken place one after another in single 20th century, which is the reason why many intellectuals had been given that thought about the world in which we are living in today [21st century] and whatever the problems we are facing is the consequences or outcome of those early back to back events, perhaps; major highlight of 20th century is United States of America had becomes a sole supreme power of the world after fall of Russians.

During and after the great depression, one person again had taken a stand to solve some of these difficult questions by writing his "General theory of employment, interest, and money" his views are still accepted and performed by many institutions including federal reserve, this revolutionary economist who challenges the neo-classical thought and argued that aggregate demand determined the overall level of economic activity and that inadequate demand could lead to prolonged periods of high unemployment, its Keynes economic thought which led the foundation of global world with highly new dynamics of economic activity with new business cycle.

There are three things, which are the essence of his thesis. We will be going to understand and analyze each of these deeply because this scenario is very much closer to our current regulatory system.

a)   The Theory of Price and wages,

b)   Investment and marginal efficiency,

c)   Macroeconomic Business model / Trade Cycle.

The Theory of Prices and Wages;

Gone are the days, when price is equal to the sum of wages, rent and profit. Earlier formulae of Price;

$$Pr = W + R + Pt$$

[46]*We move into a world where prices are governed by the quantity of money, by its income velocity, by the velocity of circulation on relatively to the volume of transaction, by hoarding, by forced savings, by inflation and by deflation.*

Such a huge things to calculate, he simplify this in one sentence that this is the theory of Individual Industry and the distribution on one hand and the theory of output and employment as a whole on the other hand.

Therefore; the modern formulae of price will be;

I. Intrinsic value (Iv),

II. Future value (Fv),

III. Investment value (In),

IV. Speculative value (Sv).

Pr = Iv + Fv + In + Sv

[47] The general price level depends partly on the rate of remuneration of the factors of production which enter into marginal cost and partly on the scale of output.

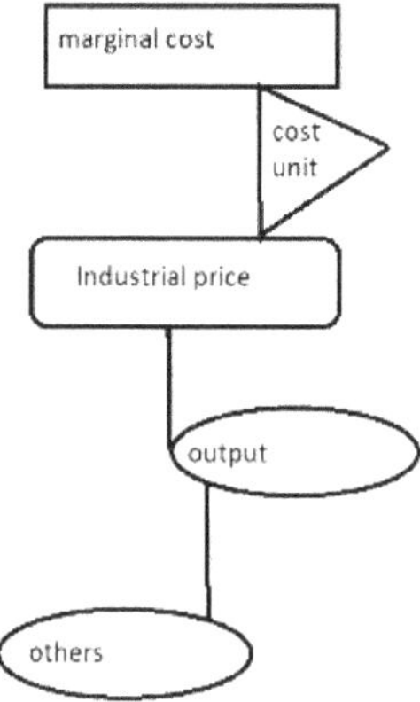

(Fig, 29; explaining relationship of price with industry)

[48] When we pass to output as a whole, the cost of production in any industry partly depend on the output of other industries.

It has to take account of demand also on both the side [cost and volume], meaning that the industry has to make its aggregate demand and supply curve, shift it accordingly to rise and fall in it and sum up the marginal efficiencies, then only it would be in a position to declare a price. But this is also true that the cost-unit is the central point of price, as he also said;

[49] And the long run stability or instability of prices will depend on the strength of the upward trend of the wage-unit [or more precisely, of the cost-unit] compared with the rate of increase in the efficiency of the productive system.

Whereas, the marginal efficiency of capital is already equals to the supply price of the rate of discount from the series of annuities offered or given by the return expected from respected capital asset.

This is the regulatory system of an industry, at this point we directly link his theory of wages here, to connect them both, we also found Keynesian thought in the favor of increased wages.

Wages

Keynes made a nice attempt on explaining the effects of money wages reduction, his conclusion helped to rebut those thought who claimed that, "A reduction in money wages will increase employment because it reduces the cost of production".

This false delusionary thought was proved wrong by his great example;

[50]*Unemployment in the US in 1932 was due either to labor obstinately refusing to accept a reduction of money-wages or to its obstinately demanding a real wages.*

He configured us to one of useful equation especially for entrepreneurs:

[51]*The volume of employment is uniquely correlated with the volume of effective demand measured in wage-unit, and that effective demand, being the sum of the expected consumption, expected investment, cannot change, if the propensity to consume, the schedule of marginal efficiency of a capital and rate of interest are all unchanged. If, without any change in these factors, the entrepreneurs want to increase employment as a whole, their proceeds will necessarily fall short of their supply price.*

$$\therefore \quad \text{employment} \quad \alpha \quad \text{aggregate demand}$$

And aggregate demand = [consumption + wage-unit + investment + marginal efficiency of capital (m.e.c) + interest rates]

If anyone of this unit got ignored, it effects the whole equation. So, for the long- run an entrepreneur can't afford to compromise anyone of this, if he wants his position at peak for long time.

Investment and Marginal efficiency;

Knowledge of investment and marginal efficiency both will helped us to understand trade cycle, its regulation, function, uses and also draw 21st century scenario of business and finance.

The definition of investment in modern thought of economics is look like;

The scale of investment depends on the relation between the rate of interest and the schedule of the marginal efficiency of capital corresponding to different scales of current investment, whilst the marginal efficiency of capital depends on the relation between the supply price of a capital asset and its prospective yield. Whereas, the total income will be find also a part of equilibrium with aggregate demand.

In Keynes words;

[52]*In other words the rate of investment will be pushed to the point on the investment demand-schedule where the marginal efficiency of capital in general is equal to the market rate of interest.*

As far as risk on investment is concerned, it's always divided on two sets but the focus point is always one, which is lender.

[53]*Two types of risk affect the volume of investment which has not commonly been distinguished.... First is the entrepreneur's or borrower's risk.... second type of risk is relevant which we may call lender's risk.*

Keynes / samuelson cross help us in figuring this game of investment;

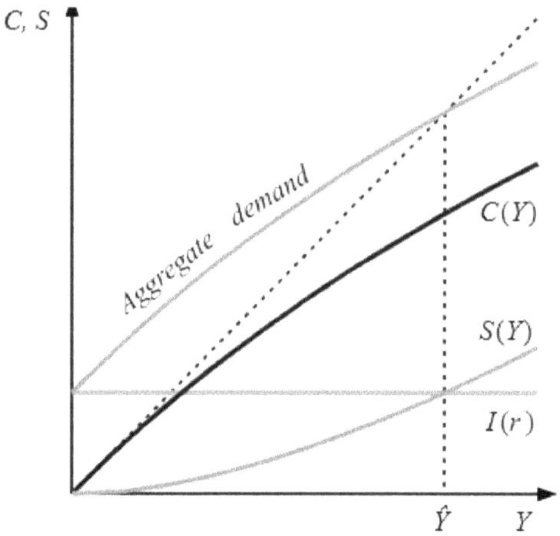

(fig, 30; Keynes/samuelson cross. P.A. Samuelson, Economics: an introductory analysis 1948 and many subsequent editions.)

Horizontal axis represents the total income, I(r);

C(y); propensity to consume,

S (y); propensity to save,

Sum of these two equal to the total income, as represents by the broken lines,

Is (r); schedule marginal efficiency of capital whose value is independent of Y.

Keynes interprets this as the demand for investment and denotes the sum of demand for consumption and investment as, "aggregated demand" plotted as a separate curve; this aggregate demand must be equal to the total income.

We make formulae of investment here;

Total income = aggregate demand ......*

Propensity to consumption + saving = marginal efficiency ......*

$\therefore$ investment = marginal efficiency + total income

Marginal Efficiency

[54]*Marginal efficiency of a capital is here defined in terms of the expectation of yield and of the current supply price of the capital asset. It depends on the rate of return expected to be obtainable on money.*

$$(m.e.c) = x1 + x2 + x3 ......$$

It is one kind of scheduled payments in where schedule is defined by loans.

[55]*The schedule of the marginal efficiency of capital may be said to govern the terms on which loanable funds are demanded for the purpose of new investment, whilst the rate of interest governs the terms on which funds are being currently supplied.*

On the portfolio of investment, marginal efficiency and their output, there is mean value of them, which is interest rates; who governs the cycle of investment.

Interest rates:

Money market can't operate without interest rates; it acts as a spinal cord in the body of economics. Keynes explained theory of interest rates with the relationship of other economic factors including monetary system.

[56]*For whilst an increase in the quantity of money may be expected, cet. par, to reduce the rate of interest, this will not happen if the liquidity preferences of the public are increasing more than the quantity of money; and whilst a decline in the rate of interest may be expected, cet. Par. To increase volume of investment, this will not happen if the schedule of the marginal efficiency of capital is falling more rapidly than the rate of interest; and whilst an increase in the volume of investment may be expected, cet. par, to increase employment this may not happen if the propensity to consume is falling off. Finally, if employment increases, prices will rise in a degree partly governed by the shapes of the physical supply functions, and partly by the liability of the wage-unit to rise in terms*

of money. And when output has increased and prices have risen, the effects of this on liquidity preference will be to increase the quantity of money necessary to maintain a given rate of interest.

r = 1/ (money + debt)

Where r; is rate of interest,

Also, there is a relation between quantity of money, liquidity preference and the rate of interest;

M = L r

Where M; is quantity of money,

L; is liquidity preference,

And r; is rate of interest.

There are six sequence of their relationship which outcomes the investment decisions with the formation of macroeconomic model.

> $M \uparrow - r \downarrow \neq M \uparrow < L \uparrow$

> $r \downarrow - In \uparrow \neq r \downarrow < m.e \downarrow$

> $In \uparrow - e \uparrow \neq c \downarrow$

> $e \uparrow - p \uparrow \cong M \uparrow$

> $P \uparrow - L \sim M \uparrow \equiv r$

> $r \uparrow - c \downarrow \sim In \downarrow$

Liquidity preference had played a role of bridge in between the quantity of money and rate of interest by fixing quantity of money on the time of holdings (by the people), at particular interest rates.

[57]*The rate of interest is not the price which brings into equilibrium the demand for resources into invest with the readiness to abstain from present consumption – it is the price which equilibrium liberates the desire to hold wealth in the form of cash with the availability quantity of cash; - which implies that if the rate of interest were lower, i.e. if the reward for parting with cash were diminished, the aggregate amount of cash which the public would wish to hold would exceed the available supply, and that if the rate of interest were raised, there would be willing to hold, if this explanation is correct, the quantity of money, is the other factor, which in conjunction with liquidity preference, determines the actual rate of interest in given circumstances.*

During the great depression, this theory works in their favor, when Keynes instructed U.S government and Federal Reserve to minimize the interest rates with the fiscal policy and invest in infrastructure by reducing taxes.

Now we are in a position to understand the gem of his economic thought:

Macroeconomic model / Trade Cycle

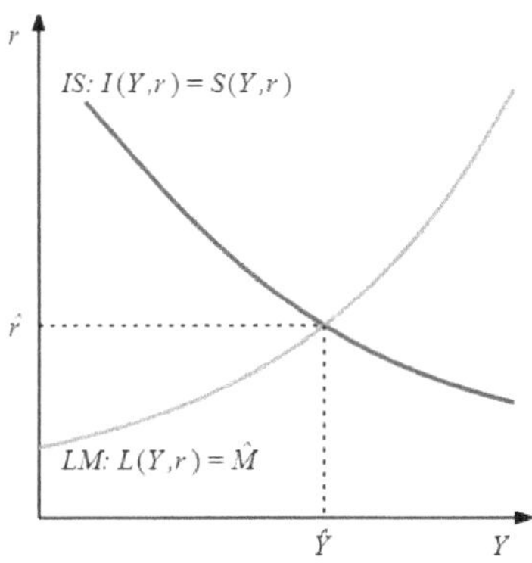

(Fig, 31; IS/LM model also called trade cycle – john hicks)

Three things came from this redefined cycle;

a)   The upward movement,

b)   The downward movement,

c)   The point of intersection.

Whereas, the point a) and b) are the two opposite curves, time is the lever of this system. At a time, when both met each other at equilibrium point, one curve get adjusted to provide a place for second curve, and when this second curve achieve its goal, it offer its acquired position for first curve again to experience the fall and losses.

In contrast to one another, those two graphical equations represented as;

    I.   IS : I (Y, r) = S (Y, r)

    II.   LM : L (Y, r) = M

First equation expresses the phenomenon of effective demand, while the second one states the equilibrium between liquidity preference and money supply. Let's say c) to the point of intersection of two curves, where total income Y and interest rate r meet each other.

Now a) in the cyclical movement travels in the upward direction, where the total income and interest rates both increases simultaneously,

b) get effected and travels in the opposite side automatically, where total income decreases but with the increase in interest rate, it loses their force.

c), the point of equilibrium comes, as a warning bell, pointing out that continuation of this cycle leads towards collapse very soon. But this warning bell gets ignored by the regulators and efforts are made to gather more forces of investment, to break this point of c), which has further advanced by the name of Hicks – Hanson model of macroeconomic, briefing the relationship between interest rates and output in the market.

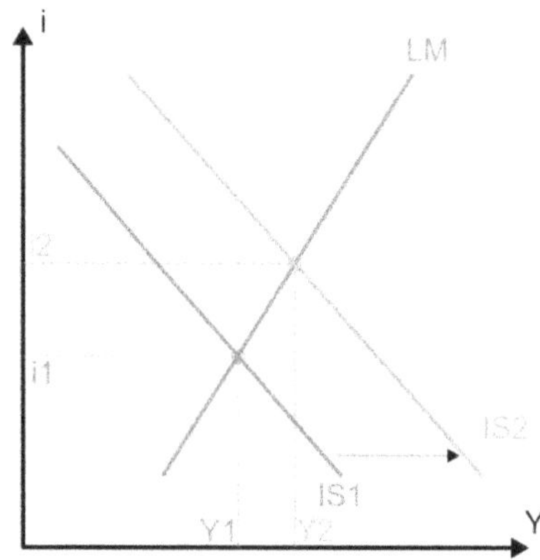

(fig, 32; post Keynesian trade cycle, shifting equilibrium state)

This mathematical calculation helps in breaking the point c) and establish new one, with the formation of new highly induced investments to reach at their destined level with the help of this mathematical adjustment a) reached to its peak and then gives his place to b) to reverse for same purpose.

Two major forces triggering this cycle and continuous fluctuations occurred respectively; these forces are Time and credit system which are directly proportional to volume of investment. Because of these two all other factors like income, savings, wage-unit, interest rates and liquidity preference with the quantity of money, they all float.

TIME;

58' The psychological time-preferences of an individual require two distinct sets of decisions to carry them out completely. The first is concerned with the aspect of time-preference which I have called the propensity to consume, determines for each individual how much of his income he will consume and how much he will reserve in some form of command over future consumption.

But this decision having been made, there is a further decision which awaits him, namely, in what form he will hold the command over future consumption which he has reserved, whether out of his current income or form previous savings. Does he want to hold it in the form of immediate, liquid command (i.e., in money or its equivalent)? Or is he prepared to part with immediate command for a specified or indefinite period, leaving it to future market conditions to determine on what terms he can, if necessary, convert deferred command over specific goods into immediate command over specific goods into immediate command over goods in general?

We shall find that the mistake in the accepted theories of the rate of interest lies in their attempting to derive the rate of interest from the first of these two constituents of psychological time-preference to the neglect of the second; and it is this neglect which we must endeavor to repair.

Brief account of time by Keynes have enlighten us that liquidity preference which conjunct interest rates and investment should need to be examine by time theory properly, because this bridge again get connected with one more hurdle of economy; credit system.

Credit system;

[59]*The notion that the creation of credit by the banking system allows investment to take place to which "no genuine saving" corresponds can only be the result of isolating one of the consequences of the increased bank credit to the exclusion of the others. If the grant of a bank credit to an entrepreneur additional to the credits already existing allows him to make an addition to current investment which would not have occurred otherwise, income will necessarily be increased and at a rate which will normally exceed the rate of increased investment. Moreover, except in conditions of full employment, there will be an increase of real income as of money-income. The people will exercise a free choice as to the proportion in which they divide their increase of income between saving and spending; and it is impossible that the intention of the entrepreneur who has borrowed in order to increase investment can become effective at a faster rate than the public decide to increase their savings. Moreover the savings which result from this decision are just as genuine as any other savings. No one can be compelled to own the additional money corresponding to the new bank. Credit, unless he deliberately prefers to hold more money rather than some other form of wealth. Yet employment,*

incomes and price cannot help moving in such a way that in the new situation someone does choose to holds the additional money.

[60]*It is also true that the grant of the bank credit will set up three tendencies;*

1. *Output to increase,*

2. *For the marginal product to rise in value, in terms of the wage-unit (which in conditions of decreasing return must necessarily accompany an increase of output) and,*

3. *For the wage unit to rise in terms of money and these tendencies are characteristics of a state of increasing output as such, and will occur just as much if the increase in output has been initiated otherwise than by an increase in bank- credit.*

These three rises are responsible for making first cyclic upward movement of trade cycle for particular time being and after the maturity of particular credit for specific time which may also called as time-preference, more authentic then liquidity preference, it started going down also termed as crisis, currently already seen in 2008 financial crisis. This trade cycle is simply trial and error doesn't fulfilled the demands for many things but at those time of 20s only available business model.

[61]*The demand for money analysis used to derive the LM curve was not based on a general equilibrium model; instead, it was assumed in a rather ad hoc fashion. It had not truly integrated the nominal and real sectors. Because it did not capture the true role of money and the financial sector, it trivialized their function. It made it seem as if a fall in the price level could bring about an equilibrium, when in fact most economists believed that a falling price level would make matters worse, not better. Nonetheless, the ISLM model was adopted. It was neat, it served its pedagogical function well, it was a rough and ready tool, it provided generally correct insight into the economy, and it was the best model available.*

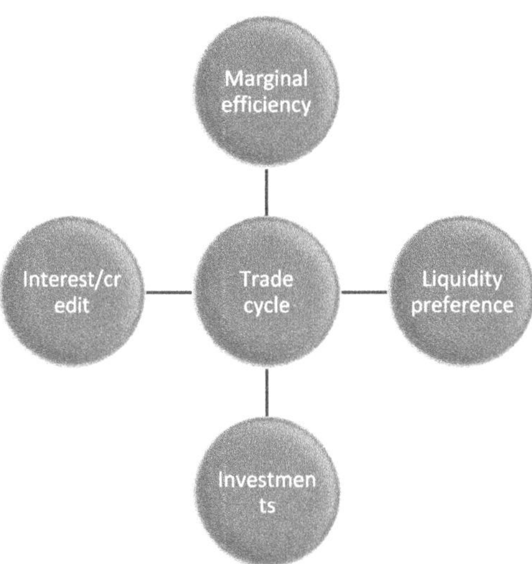

(Fig, 33; showcasing major thoughts of Keynes / modern economics)

NOTE;

In the era modern economics, where people are struggling for holding, keeping and engaging their money for their benefits, how can they think of **surplus,** the theory of surplus is again unanswered here, without surplus value neither organization nor the country will going to survive.

Third Wave of Energy and Economics

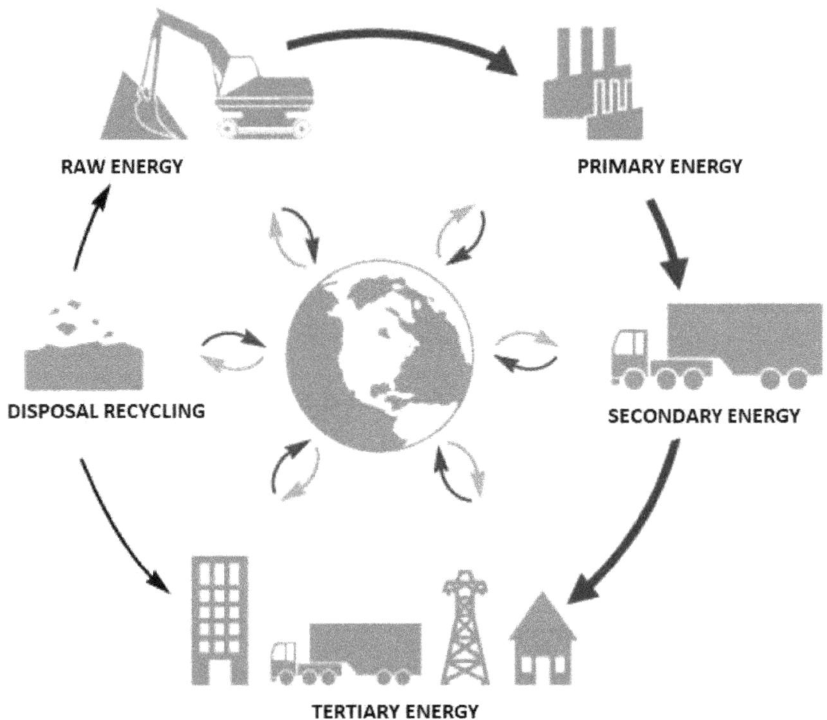

(Fig, 34; explaining the transformation of energy pattern)

This is the age of tertiary energy, the age of information and technology created by the combination of raw, primary and secondary energy pattern with its end result [-E3], which we had already obtained, Along with economic allies to make their way more easier. By the end of this wave which would have been led us to quaternary energy and the fourth industrial revolution, each individual have their personal computing system, androids, personal cell phones and all digital resources.

Cycles helped humanity to reach at this far, cycle of transformative energies, cycles of economies, changing thoughts; pre-classical, classical, neo-classical and modern economics cycles and especially the macroeconomic trade cycle with huge n huge investment make inventors of digitalization quite possible to light up this form of energy consuming by every households. These are byproducts of specific third wave of energy and economics;

- Computer's
- Television's and entertainment resources,
- Telephone's to mobile phones,
- Mass production of digital machines,
- Hyper consumerism,
- Monopoly of industries who are producing digitalization,
- Gap of rich and poor increases,
- Inflation,
- Media and press reshaping society,
- Decolonization converted to globalization,
- Biggest economic crisis followed by war's,
- Consumption of electricity increases,
- Food and beverage industry rises,
- Supply chain and logistics at peak,
- Clothing industry hike with different brands,
- Real estate's always attractive,
- Gems and jewelry new craze among people,
- Scarcity also rises with all…

How all of this had happened on one frame of time? The answer is very obvious, all production required a medium of exchange in each sector, when this medium of exchange is more larger and larger, the production rate also get larger in each zone with large outputs. This medium of exchange is money, which is running in the society like a time and makes histories with accordance to that time.

Gold Standards

Before mid-20s, gold is the medium of exchange among all; we can predict that because of gold industrialization become easy to succeed till 1914. From 1914-1925, this is the period of inflation and of course world war1, although wars couldn't be fought with gold, silver, or paper money but with the proper organization of all resources, this had been exactly happened with the proper usage of all.

(Fig, 35; picturing gold standards)

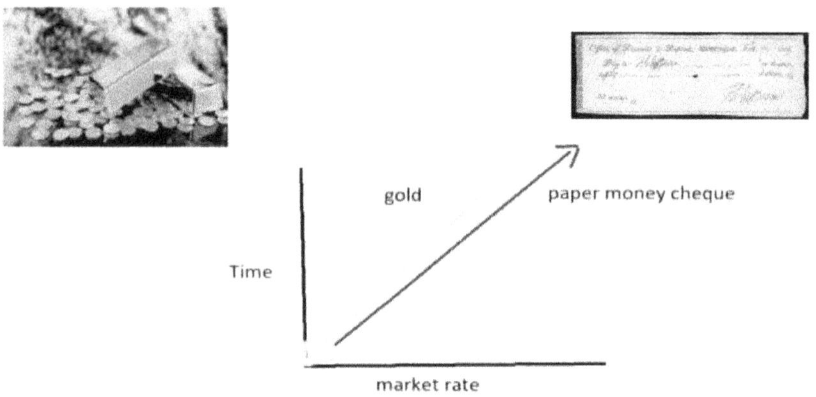

The whole scenarios of gold standards has explained in this one picture, the central bank prints paper money cheque, which are operating in the society by reserving individuals equal amount of gold, which they had asked for money, this system had made for easiness of people, because everyone was not feeling safe and secure in the times of war and each of them have to participate in daily life routine, like; traveling, buying n selling, basic needs and so on…, gold is a valuable, tangible and precious assets to have but needs security also, for this purpose every individuals of the society whether from normal households, farming, groceries to industrialists, manufacturers, traders, they all replace their saved gold to paper money cheque and using paper money cheque as the medium of exchange for their needs and wants.

The role of regulators had started at this point, the Federal Reserve has printed more paper money cheque than the saved gold reserve value, from here inflation begins with the instability of market, rise in market rate triggers the rise in demand which automatically fulfill by increase supply and for doing all of this people started buying more and more paper money cheques which transforms complete society on paper money monetary system without their awareness. This rising cycle had horrifically decline at the time of deflation.

The central bank has pulled the liver and suddenly stops the circulation of paper money cheque by announcing going back to gold again (1925-1939); people have to return those cheques to banks for their gold but the condition of exchange had changed, the risen interest rates has majorly cuts their prices and values, after long deduction people get a very small amount of their reserved gold back again in their hands and some people doesn't get any because it's all deducted, society is out of money, out of cash and out of gold too. The era of great depression subsequently invited one more phase of inflation in the year of 1939-1947, where the battle of gold and paper money continued with

Second World War, cold war's till the final diminishing of gold and circulation of paper money itself not in cheque form but in the form of valuable asset of money itself along with cheques for larger operations, from here we have the journey of our money thanks to J.M Keynes, who dismisses the old system of gold standards by this new invention with the energy of (-E3). This third wave of energy and economics has made our current world in which we are living today, with the autonomous computing, digital system and obviously with our valuable money.

Perhaps, many thinkers have admitted that previous system was far better than current monetary system;

62' The gold standard had its advantages, no doubt. Exchange rate stability made for predictable pricing in trade and reduced transaction costs, while the long-run stability of prices acted as an anchor for inflation expectations. Being on gold may also have reduced the costs of borrowing by committing governments to pursue prudent fiscal and monetary policies. The difficulty of pegging currencies to a single commodity based standard, or indeed to one another, is that policymakers are then forced to choose between free capital movements and an independent national monetary policy. They cannot have both. A currency peg can mean higher volatility in short-term interest rates, as the central bank seeks to keep the price of its money steady in terms of the peg. It can mean deflation, if the supply of the peg is constrained (as the supply of gold was relative to the demand for it in the 1870s and 1880s). And it can transmit financial crises (as happened throughout the restored gold standard after 1929). By the meaningful vestige of it was removed on 15 August 1971, the day that President Richard Nixon closed the so-called gold 'window' through which, under certain restricted circumstances, dollars could still be exchanged for gold. From that day onward, the centuries-old link between money and precious metal was broken.

3. THE CLASH

"Only the fittest will survive"

-Charles Darwin.

The universal law of nature for each time period of civilization always states that, people, product, energy, organisms, species, system and all living existing things have passed through five phases of life cycle;

a) Introduction,

b) Growth,

c) Maturity,

d) Decline.

e) Reformation.

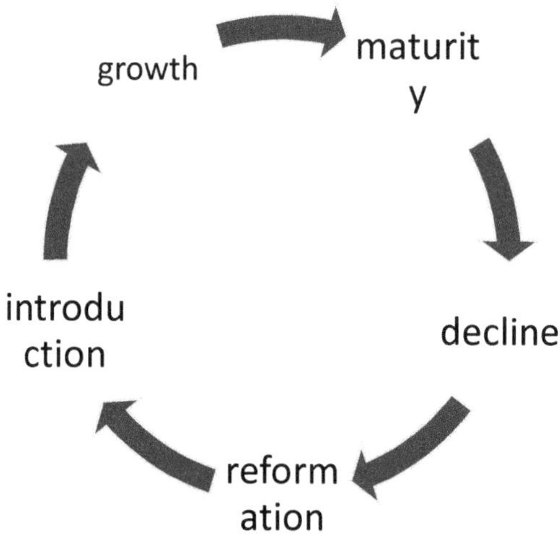

(Fig, 36 ; life cycle of living things)

It all begins with the introduction of particular subject; I will explain this cycle with the example of economics itself, firstly the subject of economics get introduced by various of economists to the people with various of thoughts, each having their pros and cons in it, after the introductory part it started growing, growth always happen after acceptance, whenever people accept some views in their day to day basis, those things has started growing just like the growth of economy had taken place, then comes the point when particular thesis about particular object gets mature, at the point of maturity great outcomes generate in various of forms, i.e. whenever capital gets mature or asset value gets mature it gives great output to investors. But then comes most unwanted part of any cycle, which is decline and also the gravitational force of our universe, in where everything which is traveling to upward direction and get place on top has to come down after their particular interval of time, just like apple fall on Newton.

On the phase of decline, i.e. after the maturity of capital or asset, when interest rates get started rising on their credit, it promotes not only the rise on debt but also decline or crisis of economy, this forth phase of cycle is always require decision making, whether we choose or adapt some new techniques and vanish the old one which led had us to this crisis or we want to continue with the same pattern again by rearranging them, in both the cases reformation has been going to occur but if first condition apply then, the reformation will be occur peacefully with the renaissance and when the second condition get placed in order to produce change, then it will definitely going to result the clashes.

The histories of clashes have shown us that it always takes place because of wrong attempts. Whenever there is exploitation, the end result is a clash, either for; civilization, power, or between; people and system, industry and labor, farmers and government, organization and employees, transportation and tariffs, energy and economics.

(Fig, 37; picturing reasons of clashes)

The formulae of clashes clearly focus on, tyranny and unjust behavior of humankind always led them to trouble with turmoil's and intolerance in the form of clashes, wars, and collapse combining low standards of living. Whereas we have already discussed in very first chapter of energy that we humankind has placed on the earth for great standards of living, bounties, success and always for victories, we are not here for clashes and uncertainty then how we can refrain from it and live peaceful life on the earth.

The footprints of clashes of history will provide complete account of their reasons, why people had been faced that? How they lacked? And why they lacked?, what are the steps we have to take for our future to avoid clashes?

As quoted by famous Dutch philosopher, Desiderius Erasmus;

"Prevention is better than Cure".

The first clash - Britain v/s Dutch

Long before Britain and its industrial power, there is Dutch who are having control of major worldly things especially economy and finance. The world's first listed company, namely The Dutch East India Company along with first stock exchange market in 1602 in Amsterdam, invented by Dutch only. They are having a very nice hold on their created capital market for the purpose of trading in domestic and overseas.

[63]*In addition to creating an equity market, the Dutch developed an innovative banking system, which grew rapidly and began to finance international trade for Dutch and non-Dutch merchants. Prior to the Dutch banking innovations, the international currency situation was a mess. In the late 1500s, around 800 different foreign and domestic coins circulated in the Netherlands, many of which were debased (i.e., had a lowered content of precious metal in the coins) and difficult to distinguish from counterfeits. This created uncertainty over the value of money, which made international trade slower and more expensive.*

[64]*In 1609, the Bank of Amsterdam was established as an exchange bank to protect commercial creditors from unreliable commodity money in general circulation. The Bank of Amsterdam undertook activities that would generate monetary stability and put the Netherland's coinage, the bank's letter of credit, and the Dutch financial system at the center of global finance. Notably this bank guilder, though backed by hard currency, was essentially type 2 money. That set up the guilders as a true reserve currency, the first of its kind.*

As a result of this system, the guilder remained effective as both a medium of exchange and a store hold of wealth. Bank of Amsterdam bills of exchange improved their status as a reserve currency. Baltic and Russian trade relied solely on guilders and bank of Amsterdam bills of exchange for pricing and contract settlement.

A series of civil war continuously seen among European countries which slowly making them weaker and weaker, Britain had used these warfare and learned closely all the do's and don'ts, these observations had helped Britain to take over Dutch in mid 18s with the dawn of industrial revolution but along with these industrial expansion, there is again a war series continue by Napoleon.

[65]*Britain's victory over Napoleon was also helped by two economic innovations: the Agricultural Revolution, which was well established there in 1720, and the Industrial Revolution, which was equally well established there by 1776, when Watt patented his steam engine. The Industrial Revolution, like the Credit Revolution, has been much misunderstood, both at the time and since. This is unfortunate, as each of these has great significance, both to advanced and to underdeveloped countries, in the twentieth century. The Industrial Revolution was accompanied by a number of incidental features, such as growth of cities through the factory system, the rapid growth of an unskilled labor supply (the proletariat), the reduction of labor to the status of a commodity in the competitive market, and the shifting of ownership of tools and equipment from laborers to a new social class of entrepreneurs. None of these constituted the essential feature of industrialism, which was, in fact, the application of nonliving power to the productive process. This application, symbolized in the steam engine and the water wheel, in the long run served to reduce or eliminate the relative significance of unskilled labor and the use of human or animal energy in the productive process (automation) and to disperse the productive process from cities, but did so, throughout, by intensifying the vital feature of the system, the use of energy from sources other than* living *bodies.*

Replacement

The rising power will always asked for their emergence with the fall and decline of existing ongoing turmoil power, which has loses all its wealth, military, people and dominance. Replacement of powers happened with the helping hands of Debt and Currency, these two factors are responsible for all clashes of human history faced and struggled by many people.

Debt in any economy act as a virus in the system, which slowly spread and poisoned it all from roots to complete body, which will without any doubt vanish that body with the name of death. And on the funeral day of that body somebody else will take a charge for future works with their implemented model and style. Indebted economy will die exactly on the same manner, whereas it will not die alone. It disappear with the disappearance of its signature, which is the current currency, using in society who allowed these virus or debt to go bigger, this currency can't be allowed to operate in market again because of the fear it created by the death of one economic system who largely indebted having this currency.

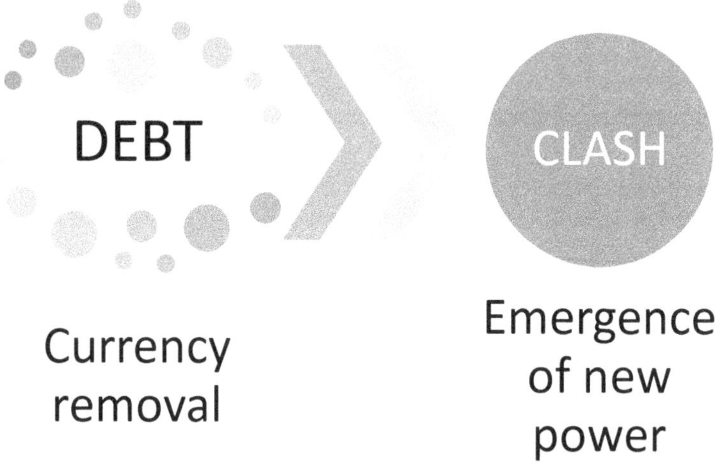

[66]*Financial losses and large debts led to the classic move by the central bank to print more money. As the bank of Amsterdam printed more and more paper money to provide loans to the Dutch East India Company, it soon became clear that there would not be enough gold and silver to cover all the paper claims on it. That led to the classic "run on the bank" dynamic, in which investors exchanged their paper money for precious metals. With the bank's store of precious metals exhausted, the supply of guilders soared, even as demand for them fell, ...*

[67]*Interest rates rose and the bank of Amsterdam had to devalue, undermining the credibility of the guilder as a store hold of wealth. As a result, the British Pound replaced the Dutch guilder as the leading reserve currency.*

The bank of Amsterdam started with type 1 monetary system (precious metal). As usual, this transition occurred at a time of financial stress and military conflict. It was risky because the transition decreased trust in the currency and added to the risk of a bank run-like dynamic, which is exactly what occurred. Bank of Amsterdam deposits (i.e., holdings of short-term debt) had been a reliable store hold of wealth for nearly two centuries. They began to trade at large discounts to guilder coins (which were made of gold and silver). The bank used its holdings of coin and precious metals (i.e., its reserves) to buy its currency on the open market to support the value of deposits, but it lacked adequate foreign currency reserves to do this indefinitely. Accounts backed by coins held at the bank plummeted from 17 million guilder in March 1780 to only 300,000 guilder in January 1783 as owner of these gold and silver coined demanded them back. The bank run marked the end of the Dutch empire and the guilder as a reserve currency. In 1791 the bank was taken over by the city of Amsterdam, and in 1795 the French revolutionary government overthrew the Dutch republic, establishing a client state in its place. After being nationalized in 1796, rendering its stock worthless, the Dutch East India Company's charter expired in 1799.

Britain won in the first clash of energy and economics, perhaps the fight is for energy itself but it seems wasted along with economy. The war between Dutch and Britishers had been fought for acquiring the power of energy but it get clashed with their economic system of credit and debt with poor management, and whenever poor economic system clashed with high, frequent, innovative energy it results the fall, and rise of opposition.

This cycle of rise and fall has seemed taken from Britain in the second clash of energy and economics by its never ending loop.

The Second Clash   -   Britain v/s colonies

(Fig, 38; picture of British Pound)

The Great Empire of British with their signature of Pound expanded with their colonies by the name of trading over demographic and geographically suited countries, among them military and navy invasion with the financial monopoly rooted British East India Company, by the help of English Rothschild Bank.

Newly monitoring currency of pound has dominated the world; more than 60 percent of the world trade was performed with the exchange of pounds globally, and made UK global trade center.

Although Great Britain had very previously started their struggle on domination of overseas, from 16 – 17th century their maritime expansion with their British American colonies in New England, Virginia, Maryland, Bermuda, Honduras, Antigua, Barbados, Nova scotia, Jamaica and Hudson's Bay Company in northwestern Canada from 1670. One of their most prominent colony which get ruled by almost 100s of year is the Indian subcontinent, comes in the predefined map of British as the conquest of Mughal Bengal at the battle of Plassey in 1757. Through Indian continents they explored pacific regions too, i.e. Singapore, Malacca, Australia, New Zealand, and from Asia-Pacific to the Cape of Good Hope.

***(Fig, 39; explaining domination of the great British empire-
highlighted countries)***

[68]*The profitability of capital outside Britain—a fact which caused the great export of capital—was matched by a profitability of labor. As a result, the flow of capital from Britain and Europe was matched by a flow of persons. Both of these served to build up non-European areas on a modified European pattern. In export of men, as in export of capital, Britain was easily first (over 20 million persons emigrating from the United Kingdom in the period 1815-1938). As a result of both, Britain became the center of*

world finance as well as the center of world commerce. The system of international financial relations, which we described earlier, was based on the system of industrial, commercial, and credit relationships which we have just described. The former thus required for its existence a very special group of circumstances—a group which could not be expected to continue forever. In addition, it required a group of secondary characteristics which were also far from permanent. Among these were the following: (1) all the countries concerned must be on the full gold standard; (2) there must be freedom from public or private interference with the domestic economy of any country; that is, prices must be free to rise and fall in accordance with the supply and demand for both goods and money; (3) there must also be free flow of international trade so that both goods and money can go without hindrance to those areas where each is most valuable; (4) the international financial economy must be organized about one center with numerous subordinate centers, so that it would be possible to cancel out international claims against one another in some clearinghouse and thus reduce the flow of gold to a minimum; (5) the flow of goods and funds in international matters should be controlled by economic factors and not be subject to political, psychological, or ideological influences.

This territorial, military and financial expansion has come to set down by the 19th century, because of their competitive rivalry, intolerance created because of inequality especially in great Indian continent, where movements for freedom also began, this huge and fat empire is very expensive for their maintenance, required a lot of capital in the form of money and labor both, this again make the chain of previously fallen Dutch Empire, large amount of debts, biggest gap of not only wealth but also politics.

The Sun Sets in the West

The sun of western empire after contributing in the rise of inequality, large debt crisis and fall of currency gets settled down by producing a tremendous shift to US which is still dominant empire of wealth and power today.

[69]*By the late 1800s, the top 1 percent of the population owned over 70 percent of all wealth, more than in peer countries. The UK's top 10 percent owned an astounding 93 percent of its wealth.*

The large gap between rich and poor and all rivalry wars against British, by their offensive colonies and political campaigning countries loaded their debts, losing their power and legal status of pounds, in addition with several devaluation.

[70]*The 1940s are frequently referred to as "crisis years" for the pound. The war required the UK to borrow immensely from its allies and colonies, and those obligations were required to be held in sterling. When the war ended, the UK could not meet its debt obligations without either raising taxes or cutting government spending, so it necessarily mandated that its debt assets (i.e., its bonds) could not be proactively sold by its former colonies. The US was anxious for the UK to restore convertibility as soon as possible, as the restrictions were reducing liquidity in the global economy, affecting the US's export profits. The bank of England was also eager to remove capital controls in order to restore the pound's role as a global trading currency, increase financial sector revenues in London, and encourage international investors to continue savings in sterling. **In 1946, an agreement was reached in which the US would provide the UK with a loan of \$3.75 billion (about 10 percent of UK GDP)** to offer a buffer against a potential run on the pound. As expected, the pound came under considerable selling pressure when partial convertibility was introduced in July 1947, and the UK and the sterling area countries turned to austerity to maintain the pound's peg to the dollar. Restrictions were imposed on the import of luxury goods, defense expenditures were slashed, dollar and gold reserves were drawn down, and agreements were made among sterling economies to not diversify their reserve holdings to the dollar.*

[71]*The devaluation came two years later, as policy makers in both the UK and the US realized that the pound couldn't return to convertibility at the current rate.*

[72]*The devaluation did not lead to a panic out of sterling, even though the fundamentals remained poor, because a very large share of UK assets was held by the US government, which was willing to take the valuation hit in order to restore convertibility, and by sterling area economies, such as India and Australia, whose currencies were pegged to the pound for political reasons. Still, the immediate post-war experience made it clear to knowledgeable observers that the pound would enjoy the same international role it had prior to World War 2.*

[73]*The UK was too indebted and too uncompetitive; it couldn't pay its debts and still buy what it needed to import. Sterling had to be devalued again in 1967. After that, even sterling area countries were unwilling to hold their reserves in pound unless the UK guaranteed their underlying value in dollars.*

[74]Countries that continued to hold a high share of their reserves in pounds after 1968 were holding de facto dollars because the sterling agreement of 1968 guaranteed 90 percent of their dollar value.

The second clash goes in favor of American's by offering them a new chance to rise in the world platform with their signature of dollar. This second meet of exploited energy and economics had produced unacceptable bloodshed on the earth; everyone seemed wanted to acquire the power, this race for power generated only wars and long lasting battle for energy and power. Along with this race, in the same parallel universe one more war of economy also ran simultaneously by following same pattern of crisis for each Empire, this chain reaction outburst many rise and fall almost for many centuries, and on this 21st century it is on its peak again.

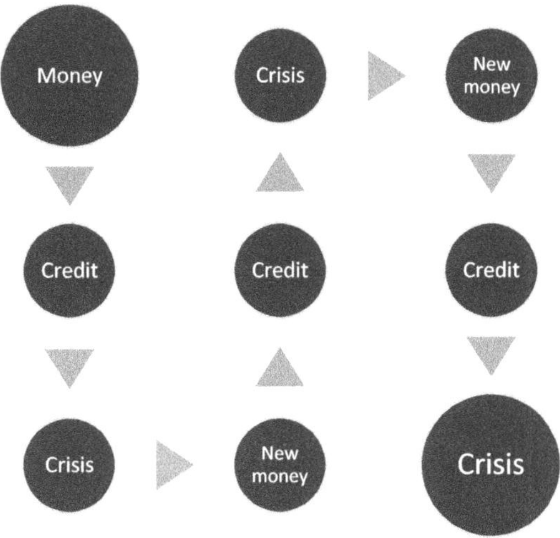

(Fig, 40; Chain reaction of money)

Third Clash - The Age of War

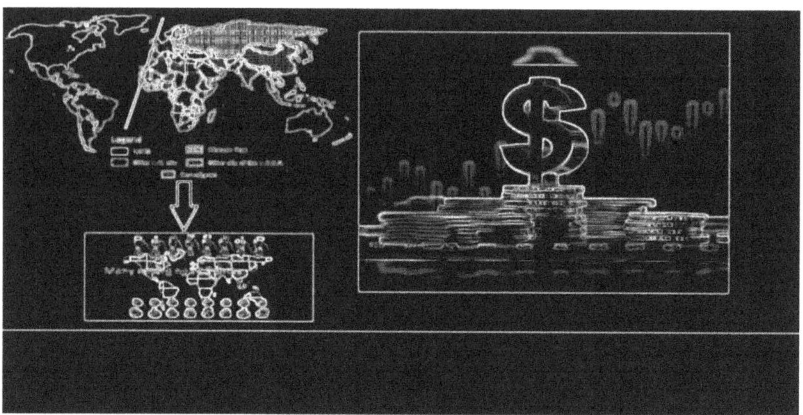

(Fig, 41; portraying world order from 20th century to 21st century)

Our Globe has been divided into two parts holding two systems, which are exactly opposite in nature with their different signature of currencies. On one side there is United States of America with its huge capitalism, dollars, economy, military and alliances, i.e., NATO, SEATO, and UN. Confronting the great USSR with its huge socialistic society, ruble, economic power, geographical benefits, and Asian–pacific control.

These two superpowers of 20th century made bipolar world at that time by experiencing and enjoying their status on changing world format, but after the Second World War and biggest historical explosion on Japan, the fall of Russians with the fall of Ottomans took place and then world has only one supreme super power running till now with their great signature of dollar which is the global currency too, all over trading was performed by dollar itself. But this domination of dollar on the world has occurred after a long series of war's and battles appeared one after another without taking any pause on 20th century to reform the next, which is our century, In our current century, we also have multipolar blocks under the supremacy of USA.

The difference between the third clash to previous two's was that on previous clashes of energy and economics, major collapse occurred on the sector of economy only with the scarcity of energy and exploitation of primary resource of energy i.e., sustainable energy. Whether in 20s the clash became horrifying for both the sectors, i.e., energy and economics.

The Third Chain Reaction

Continuing with economic factor again, here also after the fall of Great Britain the most expanded empire, United State had taken a charge because of their financial power.

[75]*As a result of the Bretton Woods Agreement, the dollar became the world's leading reserve currency. This was natural because the two World Wars had made the US the richest and most powerful country by far. By the end of World War 2 US had amassed its greatest gold/money savings ever – about two thirds of all the government – held gold/money in the world. (As* we already discussed on the chapter of gold standards).

[76]*The US did all the classic thing that helped the world become more dollarized. Its bank increased their operations and lending in foreign markets. In 1965, only 13 US banks had foreign branches. By 1970, 79 banks had them, and by 1980 nearly every major US bank had at least one foreign branch, and the total number of branches had grown to 787. Global lending boomed. However, as is also typical, a) those who proposed overdid things by operating financially imprudently while b) global competition, especially from Germany and Japan, increased. As a result, American lending and America's finances began to deteriorate as its trade surpluses disappeared.*

Third round of chain reaction started at this point, where again new money, i.e., dollar entered in the society and monetary system for beneficial operation to produce great output as input change for this purpose only. This newly arrived currency/money placed on the machine of credit to generate surplus but here again it will going to generate crisis again with the demand of new money. The same pattern had been repeated millions of time, only object had passed through change without change on the subject and operating system, currency is replacing all the time but their operation on the society is still same.

[77]*The dollar these deficits produced went to countries that were running budget surpluses, which deposited them in American banks, which lent them to Latin American and other emerging, commodity-producing countries. Savings and loans associations borrowed short to make longer-term mortgages and other loans, using the positive spread between short rates (which they borrowed at) and long rates (which they lent at) as a source of and its effects on markets came in two big waves that were*

bracketed by periods of extreme monetary tightness, steep stock market declines, and deep recessions. Early in the 1970s, most Americans had never experienced inflation, so they weren't wary of it and allowed it to blossom. By the end of the decade, they were traumatized by it and assumed that it would never go away.

[78]*Throughout the long-term debt cycle, from 1945 until 2008, whenever the Federal Reserve wanted the economy to pick up it would lower interest rates and make money and credit more available, which would increase stock and bond prices and increase demand. That was how it was done until 2008- i.e., interest rates were cut, and debts were increased faster than incomes to create unsustainable bubbles. That changed when the bubble burst in 2008 and interest rates hit 0 percent for the first time since the Great Depression.*

[79]*Central banks printed money and bought financial assets, which put money in the hands of investors who bought other financial assets, which caused financial assets prices to rise, which was helpful for the economy and particularly beneficial to those who were rich enough to own financial assets, so it increased the wealth gap. Basically borrowed money was essentially free, so investment borrowers and corporate borrowers took advantage of this to get and used it to make purchases that drove stock prices and corporate profits up. This money did not trickle down proportionately, so wealth and income gaps grow. Wealth and income gaps grew to the largest since the 1930-45 period.*

This running reaction has seems to be broken on coming years, at the timeline of fourth industrial revolution whom created its owned signature by itself without the help of any Nations, it is digital currency.

-E3

Global primary energy crisis had firstly begun with the Industrial revolution by the rise of coal, steel and biomass.

$$E \;\rightarrow\; E1$$

Then ranges to fossil fuel, oil, gas then hydropower plants, consumption of increased electricity and mass production again leads to scarcity and famine of primary resource of renewable energy, till the time of world wars. After post war era, during cold war and civil wars, two biggest energy crises occurred; 1) on 1973, OPEC by Arab member states, and 2) on 1979 as a result of Iranian revolution.

Both of these major crises happened for the same thing, oil. The backbone of modern nuclear power Century flooded the wave of energy to the level of –E3.

Energy Information Administration, represented energy review in 2006 which forecasts oil crises, per barrel dollar.

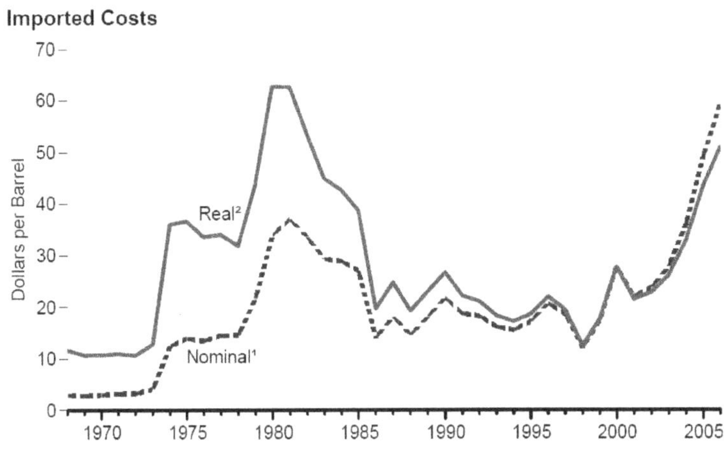

(Fig, 41; third clash of energy)

The negative sign of energy has started appearing with maturity whenever the consumption of transformed energy (non-renewable) becomes greater. The greater unsustainable consumption has negatively impact on sustainable and natural energy, (-E).

On opposite side, the lesser or balanced consumption of non-renewables has positive impact on natural energy which again help in many kinds of transformations.

Global primary energy consumption by source

Primary energy¹ is based on the substitution method² and measured in terawatt-hours³.

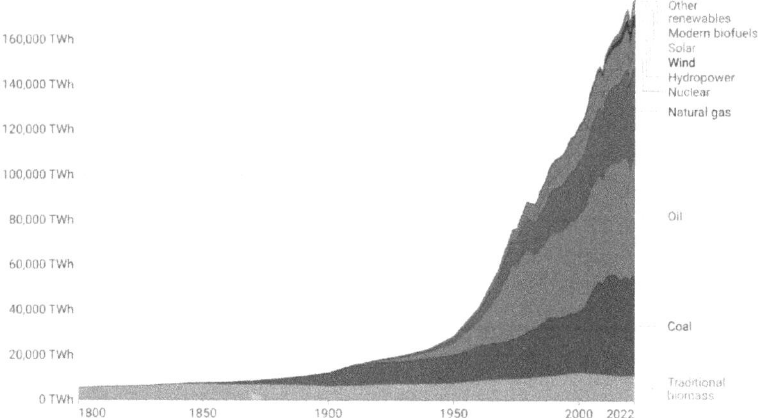

Data source: Energy Institute - Statistical Review of World Energy (2023); Smil (2017)   OurWorldInData.org/energy | CC BY
Note: In the absence of more recent data, traditional biomass is assumed constant since 2015.

1. **Primary energy:** Primary energy is the energy available as resources – such as the fuels burnt in power plants – before it has been transformed. This relates to the coal before it has been burned, the uranium, or the barrels of oil. Primary energy includes energy that the end user needs, in the form of electricity, transport and heating, plus inefficiencies and energy that is lost when raw resources are transformed into a usable form. You can read more on the different ways of measuring energy in our article.

2. **Substitution method:** The 'substitution method' is used by researchers to correct primary energy consumption for efficiency losses experienced by fossil fuels. It tries to adjust non-fossil energy sources to the inputs that would be needed if it was generated from fossil fuels. It assumes that wind and solar electricity is as inefficient as coal or gas. To do this, energy generation from non-fossil sources are divided by a standard 'thermal efficiency factor' – typically around 0.4 Nuclear power is also adjusted despite it also experiencing thermal losses in a power plant. Since it's reported in terms of electricity output, we need to do this adjustment to calculate its equivalent input value. You can read more about this adjustment in our article.

3. **Watt-hour:** A watt-hour is the energy delivered by one watt of power for one hour. Since one watt is equivalent to one joule per second, a watt-hour is equivalent to 3600 joules of energy. Metric prefixes are used for multiples of the unit, usually: - kilowatt-hours (kWh), or a thousand watt-hours - Megawatt-hours (MWh), or a million watt-hours - Gigawatt-hours (GWh), or a billion watt-hours - Terawatt-hours (TWh), or a trillion watt-hours.

(Fig, 42; changing pattern of energies)

$$OIL \;>\; COAL \;>\; NATURAL\ GAS \quad \cong \quad Non\ Renewable\ Energy$$

$$HYDROPOWER \;>\; BIOMASS \;>\; WIND \;>\; SOLAR \;>$$
$$OTHER\ RENEWABLE \quad \cong \quad RENEWABLE\ ENERGY$$

Summary of Book 2

✓ Everything which is created in the universe has specific function and purpose,

✓ Nature and humans both are dependent on each other for making peace on Earth,

✓ Each civilization has passed with the efforts of survivals, position and formation,

✓ Each century had framed so many contributors in various of sectors,

✓ Everything is made up of energy and acquiring of energies require financing which in combination makes economy, when society wins in both it has political power when lose it has clashes and wars,

✓ All economists of their time had came up with his theory to acquire more and more natural resources by using different of methods,

✓ Smith came with capitalism, Marx with socialism, Marshall with mathematics and Keynes's with cycles;

✓ All have same factors of production namely; land, labor, capital and entrepreneurship/organization,

✓ Again these four has subdivided by their unsolved answers;

- Land- has issue of rent,

- Labor- has issue of wage,

- Capital- has issue of credit, interest rates and profit,

- Organization- has issue of pricing, revenue, income-expenses.

✓ Meanwhile energy sector during these economic evolution transforms through zero to peak,

✓ Last four centuries has resulted four revolution back to back in each century,

✓ Each revolution has a cycle; base – incline – decline – reform,

✓ At the phase of decline clashes appear upon societies with tragic losses,

- ✓ Humanity has already gone through three major clashes of history, which minimized natural resources and damage, perhaps maximized non – renewable usages,

- ✓ At this point, nature is demanding shifting of systems, previous methods expired very long it can't be useful in coming world which is going to be global world, previous techniques if get applied in global Way it will definitely results the biggest clash of clashes in the history of energy and economics.

BOOK – 2

THE 4TH QUADRANT

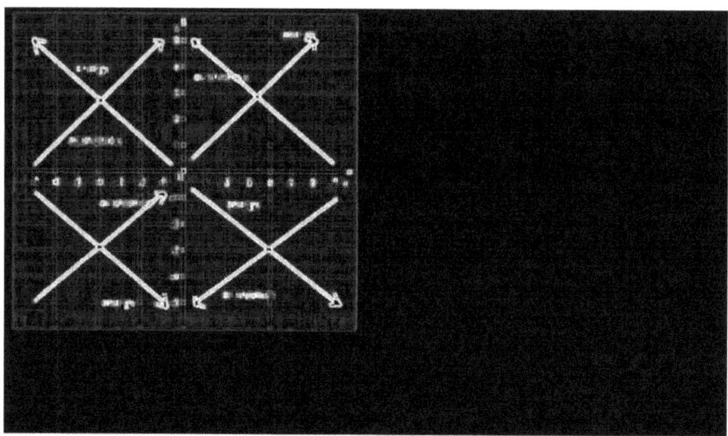

(Fig, 43 ; explaining four biggest era with four parameters)

On x-axis; money/investments,

On y-axis; debt/credit.

Energy and economics both are running on opposite direction.

The journey of energy, economics, money market and their investments at credit on borrowed debt, they all are seemed to be growing very rapidly each other in the very first quadrant, where we easily seen;

Energy	Economics		
Money	↑	>	↓
Debt	↑	≅	↑

The whole economic system with their huge investments and complete monetization engaged to increase energy output with the rise in capital involved in energy while decreases in money market of economy. This input towards energy by economics also has a major byproduct, present in all the times. I.e., credit or debt, because all monetary system is based on interest – theory, which promotes continue rise in credit and debt in both productive sector of energy and economics, where debt is parallel on both the side each sector has rising overheads of debt.

Energy || Economics

The parallel situation occurred only in the case of debt and credit.

Indeed, this first chamber is the witness of only rise, in all sects of life.

Then, comes the second quadrant with one minus value on x-axis which is money/capital/investments and other plus sign on y-axis, which is credit and debt.

Energy		Economics		
Money	$-\uparrow$	$>$	$-\downarrow$
Debt	$\uparrow$	$\cong$	$\uparrow$

Same pattern of debt at same level of increasing power to certain value have shown, while negative increasing money in energy sector and negatively decreasing money in put economics which lead the decline of Dutch and their gliders with the rise of British in second quadrant with their pounds, who already indebted before coming monetary system [as minus sign showing their money value].

Parallel of debt found again in both sectors with their maturity (+ sign),

Energy || Economics

The worst pattern occurred in third quadrant, where all four components of system negatively up sided downed. The biggest collapse of economy with all whole market including money, credit, investment, stocks and bonds, they all fall with energy sectors in their both resources; renewable and non-renewable. In which nuclear and oil crisis was one of the major parts of this third segment.

Energy		Economics		
Money	$-\downarrow$	$\sim$	$-\downarrow$
Debt	$-\downarrow$	$\sim$	$-\downarrow$

Previously risen debt on both first and second quadrants has now negatively fallen down by crashing all the system of energy and economics on which they all are reliable.

All these recurring events had led human civilization on the path of new world order, where humanity needs a new organized world, full of easiness, freedom and better place to live and here we are, on the way of coming mid – 21st century and our fourth quadrant.

Energy || economics || money || debt

The Fourth Quadrant

"I feel like the smartest people in my field are busy reinforcing the old models with the new technology".

- Douglas Rushkoff.

The new beginning of all the factors has started appearing;

	Energy		Economics
	Energy		Economics
Money	↑	>	↓
Debt	−↑	≅	−↑

Exactly opposite angle of 1st quadrant going to happen in 4th quadrant of our system, where debt rate on both the side of energy and economics negatively risen very high at same rate and same value, while the derivative of this credit, i.e. money will find in favor of energy again, it seems to continuous increase with continuous decrease in economy from high rate to low and vice versa.

Debt on energy and economics is again running parallel negatively.

Energy	‖	Economics

But why we have this debt throughout this last quadrant on both the sides, well the answer will not only amaze you but also shook you.

But we people of this ongoing century have to understand this quadrant, because our lives are now dependent on this 4th quadrant, which will be going to make our future ahead.

This fourth quadrant is nothing but the shadow of fourth Industrial Revolutionary body which is under our experience, also labeled as the era of Artificial Intelligence where energy will be seen on the place, where no one can imagine its height, use and advanced form.

This will be going to very big form of unsustainable energy by the help of machines, robotics and all artificial driven forces to our lives. Whether you consider it as positive or negative, doesn't matter that now, the only thing which matter here is only those will going to survive on this quadrant of system who are running with this system, but not blindly, perhaps wisely and the wise demand of that system is not energy but clearly defined economics.

We are rapidly growing in energy but as economy is concern, we still on the back of modern economic thought which was already died before the creation of newly designed world order. Here on the progressively changing world factors we people have to put our attentions to developing new thoughts of economy for their perfect match with this highly energized time, why will economic thought be constant when everything is changing, it's the right time to invent Global Economic Thought for Global World.

But before going to remake this Global Economic Thought, we have to reconsider the developing fourth industrial revolution, in our chapter of fourth wave of energy and economics and also its consequences, in our chapter of fourth clash of energy and economics. The optimistic data analysis of this combined fourth quadrant will quickly help us to frame our economic thought with blueprint for our future ships which can't sink.

The Fourth Wave Of Energy And Economics

"Generative AI: Steam Engine of the Fourth Industrial Revolution?"

-World Economic Forum, Annual meeting, Davos 2024.

According to reports, the environment of 20s has drastically changed the natural system of earth by the fourth wave of energy and economics, in where multi-layer change occurs;

80' The unparalleled wealth generated by nearly three centuries of industrialization has not only been unevenly spread among people, but has come at a significant cost to Earth's natural system; climate, water, air, biodiversity, forests and oceans are all under unprecedented, sever e and increasing stress. Species are going extinct at up to 100 times natural levels. In 1800, only 3% of the world's population of 1 billion lived in urban areas. Today, more than 50% of the world's 7.4 billion inhabitants live in urban areas. Of these, more than 92% experience air pollution above levels deemed safe by the World Health Organization. By 2050, there will be more plastic than fish in the ocean by weight. Global CO2 emissions have risen 150-fold since 1850. And at the current rate of emissions, the risk that the world will be between 4ºC and 6ºC warmer by 2100 than it is today is real, which could irreversibly alter the otherwise stable climate system that we have enjoyed for the last 10,000 years.

And in correspondence with these environmental issues, there is more critical change on the wave of energy and economics itself, where the wave of both gets unified by the networking of digital records. Phrases as Block chain, this fourth wave is in reality block chain wave of both energy and economics where we will witness them as one unit working together where none can differ or distinguish from other, both get collided.

[81]This is revolutionary for four reasons. First, block chain technology helps overcome the double-edged sword of the digital economy _ the fact that digital objects can be copied exactly and transmitted at almost no marginal cost to multiple people simultaneously. This is valuable for sharing information but is problematic when transmitting something of unique value or guaranteed provenance _ whether a unit of a digital currency, a document that contains indispensable information or perhaps a piece of art, where knowing who holds the original is important. Block

chain enable the creation and transfer of verifiably unique digital objects, without the risk of false copies or double-sending, creating what has been called "the internet of value".

The second revolutionary aspect is that distributed ledger technologies allow transparency, verification and "immutability" without requiring anyone to trust a single central third party. This is important because situations abound where it is extremely difficult to trust, agree on or set up a third party to record the details of transactions, or assert the source or ownership of a valuable asset.

The third important attribute is that distributed ledger allow for programmable actions – transactions that can be executed (and then traced and verified) without human intervention. This ability goes beyond algorithmic trading or automated online transfers. Smart contracts on a block chain can be designed to transfer any piece of information or assert under any set of specific circumstances, from an insurance contract that pays out when rainfall levels exceed a certain amount, to automatically distributing royalties or rewarding multiple parties for different amounts of work on a project. Importantly, the code that executes the smart contract is in itself stored on the block chain, is available for inspection and runs for everyone without delays.

Fourth, digital ledgers can be designed to be inclusive. Block chain transactions are by nature simultaneously transparent, secure and traceable. If desired, they can also be anonymous. At least for the user, making a transaction requires little bandwidth and requires only basic software, storage and connectivity. This means that individuals and small contributors who normally would be excluded from markets can become market players as producers, shareholders, beneficiaries or consumers of any asset capable of being tracked and traded in a digital form.

The economy of block chain has seemed to be completely dependent upon Electrical Energy,

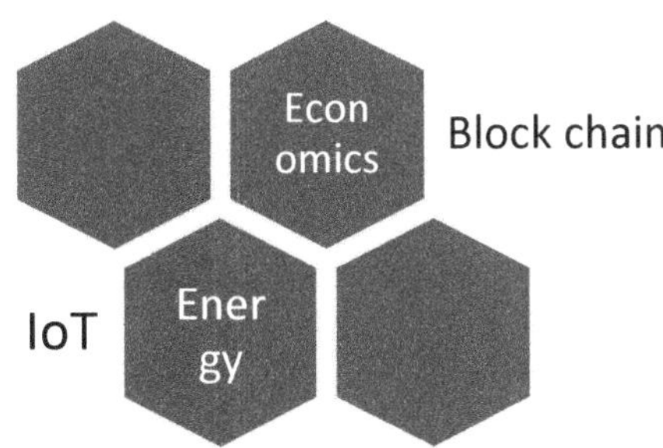

(Fig, 44; explains unification of waves)

The block chain of economic system, melted with IoT (Internet of things), just like the melted gold from which you can't pick the original metal again. Somehow this also has done right that way;

[82]*First, it enables rich data to be combined with smart analytics, which provides new sources of contextual data reflecting events in the wider environment. It also provides device performance data, helping firms and individuals anticipate how assets are performing and where opportunities to extend value exist. It will also deliver user – impact data, showing the effects of how, when and why people take actions. This enabling capability will reshape what we know and prioritize how we make decisions.*

The second core capability comes from these devices communicating and coordinating in ways that enhance efficiency and productivity. Both end to end automation and new forms of human – machine collaboration will streamline routine tasks and enhance individual's ability to apply creativity and problem – solving skills to higher – value challenges. The ability to expand from an administrative and task – oriented mindset can shape more synthetic perspectives as people become accustomed to considering peripheral input in the shaping of products, services and ideas.

The third capability is the creation of intelligent – interactive objects that provide new channels for delivering value to citizens. As a distributed network of sensors and devices, synergistic opportunities exist for other distributed technologies, such as cloud AI, block chain, additive manufacturing, drones, energy production, and more. With these new technologies converging, the decentralization of value creation and exchange will mimic the infrastructure that enables it, and the outcomes of this economic reformatting are likely to surprise us. For this reason, IoT will ultimately challenge existing institutions and conceptual frameworks on how to think about the nature of products, services and data, as well as how to think about the definitions of their value in a way that works for business.

Thus, these three capabilities will create the impetus for changes to business models and structural shifts across a wide range of industries, including manufacturing, oil and gas, agriculture, mining, transportation and healthcare.

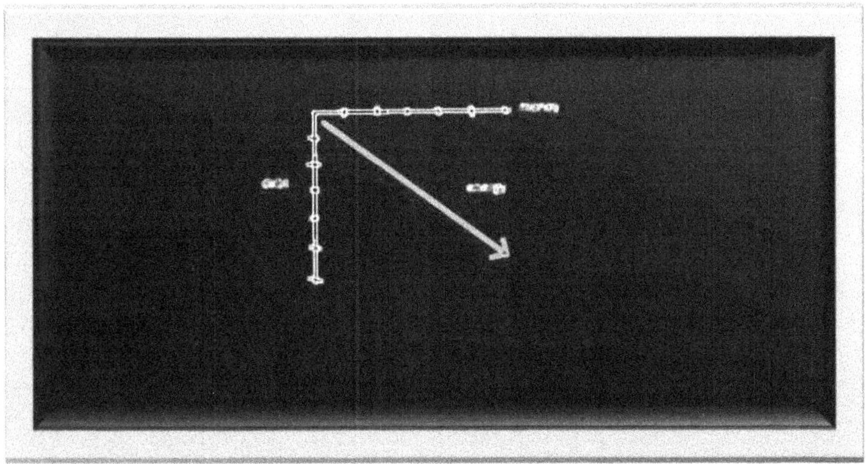

(Fig, 45; explains 4th quadrant energy of AI)

Money and debt on Artificial Intelligence runs equivalently, where investments push machines, data and cloud computing with the negative debt rate.

It starts with point, (1,-1);

End goes to point, (5,-5).

"4IR", the strongest wave of energy and economics in collaboration of both, where they both are directly-dependent on each other by one single factor, which is Electricity. All operations as we find in the writings of Schwab and world economic forum are performed on the platform of electricity, without this electric power nothing can be done neither for digital one world, nor for machineries. Also risk of cyber security rises with electricity.

There are also some byproducts came out from this fourth wave, as small fishes thrown by oceanic waves. i.e.

- ❖ Labor replaced by machines,

- ❖ Employment of humans replaced by employment of robotics,

- ❖ Wages theory not required anymore in economy, because man power has replaced by machine power,

- ❖ From land (agriculture) to industries (organization) system will be completely on automated pilot mode,

- ❖ More burnings of fossil fuels and carbon emission,

- ❖ Mining, gas and other non-renewable resources have high consumptions, but only for some hands,

- ❖ Geo-engineering and biotechnology gain full attention.

- ❖ Money market will completely internet market.

Perhaps, from bottom to neck every product of goods will be chained to electricity, as well as the data of world population in future clearly shows upward increase, which is projected to reach 8.5 billion in 2030, further increases to 9.7 billion in 2050, and lastly 10.4 billion on the next century which is 2100. Rising population will also have a mass demand of this sacred electricity to be in a system because the whole system will unified on one single cell, i.e. Electricity.

The electrical energy will either lead this wave or fall to tsunami.

The Fourth Clash; Catastrophic Clash

"The future of money is digital currency"

-Bill Gates.

Yes, indeed this gen-z generation is going to play with digital money, where money from your hand will totally vanish and it will be only seen on your screen {crypto currency, bit coin, other forms of digital currency}. On the basis of this digital money, which will highly injected in both energy and economic sectors artificially will automatically incline debt rate at very faster rate than previous time, this sharped, negative debt rate will results in the biggest clash of energy and economics to humankind which history couldn't have witnessed.

Our previously discussed chain reaction of money, which in turns to credit, follows crisis to the new invention of money, has also applied here by using exact formulas and patterns.

United States with their signature of dollars taken British Empire, and restarted chain reaction; dollars entered in money market, huge investments in energy sectors bombarded the money and credit flow with the name of capitalistic society. This newly invented money has started losing their value with the increase in debt rate, at last balancing itself with numerous fluctuations but no one knows till when it will survive with over – burdened debt, i.e. currently USA alone have more than $30 trillion debt leaving far from its economy.

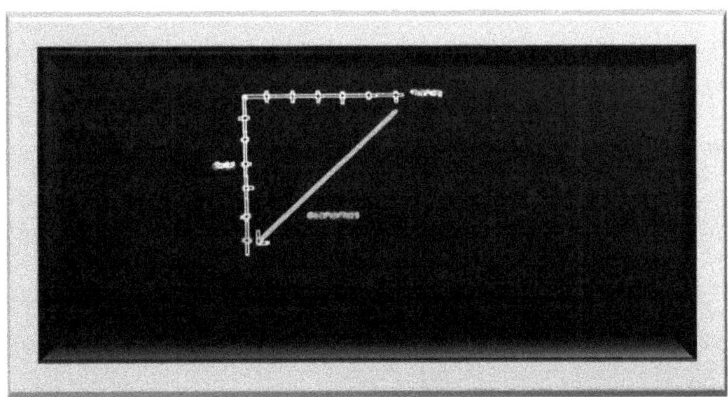

(Fig, 46; showing 4th quadrant graph of dollar and economy)

Economy with the currency of dollar at rate of (5, -1),

Where point 5 is – money,

Point -1 is – debt,

It ends on, (1, -5),

Where point 1 is – money,

And point -5 is – debt.

These high debts can't repay, inflation rate will also shake the global platform and again the regulators will get the chance to introduce new kind of currency to world for monetization and that will be the turn of digital currency, time for crypto, after the crisis of dollar.

Crypto – currency will take over the whole system of economy with their frequency of long-term investment portfolio which allows Artificial Intelligence and machine learning to take their time for their end results.

The clash in economic sector will find with the collapse of dollar to the beginning of one mutual currency to all, whereas the collapse of this digital currency will be catastrophic. And this Big Crunch will feature because of complete shutdown of Electricity.

The electricity will highly consume on this fourth quadrant, from domestic use of individuals to all economic, energy, healthcare, logistics and every zone which is responsible for living. All these chambers of our lives knotted with oneness of system and this will unanimously cuts down everything.

There is one more kind of investments which is still continuous like legacy and that is never ending wars between innocents and tyrant's on the earth. This journey of war which is started with the invention of iron, formerly with swords to now robots, always promoted by those who have power of both economy and energy, more specifically the power of money, but their power will also became the reason of their fall, because of their exploitation of money in the form of debt for supporting the warfare. The exact same scenario has find in United States, their debt market getting quadruple during the times of their unconditional support either ongoing Russia – Ukraine war or continuous Israel campaigns in middle – eastern countries.

These ongoing battles not going to stop just here only, it is contagious. The outcomes of fourth industrial revolutions would be;

- Inequality,

- Climate change,

- Urbanizations,

- Deforestation,

- Natural disasters,

- Long-term investments and long-term debts,

- Recessions,

- Unemployment's,

- Biggest gap of rich and poor,

- Civil wars,

- Poverty,

- Wealth only among 10%, remaining 90% live of low human index

Unfavorable conditions on earth increases intolerance, continental clashes will diminish allies and their breakings of treaties will reproduce global wars.

"Total war is no longer war waged by all members of one national community against all those of another. It is total … because it may well involve the whole world".

-Jean – Paul Sartre.

The Global Economic Thought

"The significant problems we face cannot be solved at the same level of thinking we were at when we created them"

-Albert Einstein.

What is meant by Global?

Anything which comprises the whole is termed as global, e.g. the universe in which we live in is global, we all human being globally living there, the concept of globalization sounds new on this century whereas it was the very early phenomenon since the creation of universe. The exact same meaning of globalization has also applied to us also. Human body is also a global thing, our whole body with their vital organs globalized to perform all the tasks for the survival of body.

This philosophical concept is now going to apply on the system too, as it applies on universe and humans, for whom universe has created to the system which maintain the balance of both; the universe and humans.

But before we jump to discuss globalization of system, specifically globalization of economic thought. We have to get the key of globalizing power, the key of globalization is always around us if you don't believe me then look around you and look in you. We all are made up of compound effect of globalization, everything around us are existed in the form of pairs; plants, animals, humans, even planets. Each and every functioning thing on this plane of universe are operating in pair and this pairing has make globalization more beautiful from the day Heavens get paired with their galaxies and planets, Earth get paired with its natural resources, like ocean, mountains and forest, and most importantly pairing of Humans the reason of Civilization.

Equivalently, we also are putting these facts on our global economic system to make it accurately natural and error-free for long term projects and developments.

The key for global economic thought is pairing, (as discussed by reasoning), now the question arises here is that what we are going to pair? And how we are pairing it for better results?

We are going to pair energy and economics (E), by making their orbits co-lateral, both of them will function by the spirit of money/capital and efficiency. As far as How is concerned, for better results; this pair of (E) if not going to apply blindly with idle approach, whether it will correspondingly paired with economic organs of the society.

$$E \ = \ me2$$

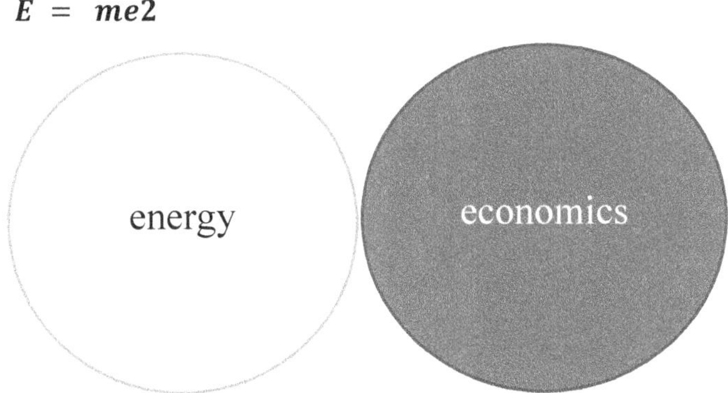

(Fig, 47; basic formulae for global economic thought)

Where E stands for both; E- Energy and Economics,

And each get calculated by their spirit which make their functioning easy, i.e. money, the invested capital among them for production, but this production will be taken place efficiently with the help of productive efficiency along with allocation of equity.

$$\therefore \ m - money \ ;$$

$$e2 - efficicency \ and \ equity.$$

The equity financing for global economy will take the whole system towards allocative efficiency where price doesn't clash with the quantity demand and supply. Also, a producer has to ship only efficient products to consumers at the rate of marginal benefit and consumer also take his unit at the rate of marginal cost. International market of prices will resolve by alliance of consumers and producers, in addition to efficiency in both the sector of energy and economics will also produce positive results by less wasting of energy, time, resources and scheduled payments.

Equity with efficiency will injected on the pair of (E), now this highly injected atom of energy and economics are profoundly ready to vaccinate major economic organs. This vaccination will firstly repair those organs, cure them and after the cure it will boost them for multiple developments. These major economic organs are;

1. Trade,

2. Infrastructure,

3. Agriculture,

4. Manufacturing,

5. Entrepreneurship.

These five parts of economic body will sufficient enough for global impact, even though this pentagon has make global body on which economic thoughts get applied with global dynamics.

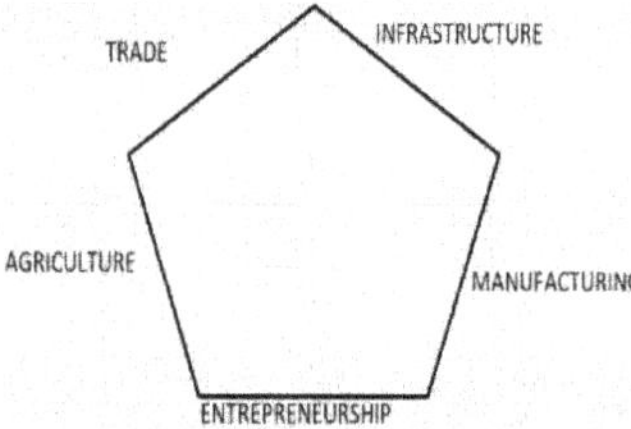

This pentagon fueled by the pair of (E), which helps in their regulations as same as happen in the human body, where all the organs automated by the supply of blood. Role of money in the economy is also like the supply of blood, when money with its true nature get supplied to these five major organs of economic body, it will nurture the society. And the true nature of money can't be finding in the form of debts, because debt always demands the repayments of principal amount with some additional multiplied amounts, these additionally charged values will lose natural values of money by exploiting its intrinsic and extrinsic value.

Again, take the example of blood, is any organ of the body has to repay more and more blood to any other organ, well of course not and it is the against of nature too, it will damage the whole system of body, in fact non rhythmic flow of blood with inequality and inefficiency will promote cancer.

The cancer of economics too has been seen throughout history of financial and economic system, recently on the beginning 20s also we all witnessed credit – default swap and their big bubbles. Cure of this dangerous disease will now apply globally for the global relief, which will be possible only when money use and reuse it in on a loop with proper efficiency by maintaining its true nature which gives fruitful surplus to all sectors of economies and societies.

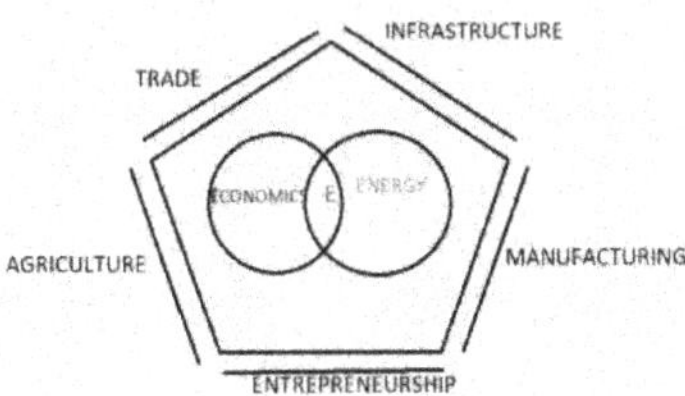

The nuclear fission of (E) planted on a pentagonal organs of global body has to be covered with graphene shield to protect entire system of global economy. This graphene shield is actually the core and pillars of our economic thoughts, on these foundations only we are going to led the global economic thought which will going to transform our system completely.

These basic fundamentals are;

1] Psychology,

2] Technicality,

3] Demand,

4] Supply,

5] The invisible hand of Price,

6] Surplus.

As we all know that we are living in the age of Machines and in future, machines will be in a position of new creatures made by man will programmed to live with man even in more standard and sophisticated life style compared to man itself.

Considering high rank machines in future, it is the demand of time that we also have some kind of global economic machine which will not only help us against drowning now but also in coming times too. For accomplishing this demand I have made Global Economic Machine, which is not like regular economic cycles but the economic base.

The Global Economic Machine

(Fig, 49; Global Economic Machine)

1] Psychology;

Behavioral economics with the subject of study of man, where we examine individual behaviors under different circumstances, transformed this examination to predictory analysis and statistics results some time inflation, sometime deflation. An overall result is not so impressive and doesn't even seem surviving in the long run.

Perhaps, economics is not the subject of study of man rather it is the study of psychology. In our coming world where humans have to survive in accordance to machines, they have to compete with these robotic machines that have more cognitive power than group of individuals, on those circumstances leadership demands understanding of psychology.

Psychology of both; humans and machines must be a center point for global economy and has absolute different nature with each other. Starting with the basic definition of cognitive psychology which states that,

[83] *"Cognitive psychology is the scientific study of mental processes such as attention, language use, memory, perception, problem solving, creativity, and reasoning."*

According to this definition human psychology has been built by a long process of their upbringing, education, social circles, professions, activities and many other learning experiences which all required various time periods. Perhaps, the psychology of machines and robotics are just far away of one click, within a few seconds you can transfer all the readily available data to machine by their programming and your completely organized and well programmed product is ready.

So, here we have to understand their psychological responses on various of coming events, as for humans who have different psychologies according to their background, will may take some time to analyze their response and to accumulate them based on their expertise. Whereas for machines and robotics it will be very easy, because we are the only programmer of them and we will be very much aware of their response, their expertise and their faults.

This is why employment of humans has becomes a major global concern. The biggest threat for people's survival is not artificial intelligence but the conscious intelligence which can be unbeatable. There is one thing for sure is that we can't defeat the machine on the data base, cognitively they are much superior than us, still people are trying to challenge it by adapting machinery qualities on their selves;

[84] *"Neurotechnologies enable us to better influence consciousness and thought and to understand many activities of the brain. They include decoding what we are thinking in fine levels of the detail through new chemicals and interventions that can influence our brains to correct for errors or enhance functionality. They also help us find new ways to communicate and interact with the world, as well as opportunities to dramatically expand our senses."*

[85] *"The brain works through electrical signals initiated by chemical interactions; these can be measured, desirable signals mimicked and undesirable ones prevented from propagating through the brain, all by influencing either brain chemistry or electrical signaling, specialized technologies, such as modern microelectrodes, can record the activity of a single neuron, or trigger it as needed. Functional magnetic resonance imaging can reveal how different regions of the brain are active in different circumstances."*

Conceivably neurotechnologies grant seviour health related issues, which is again unacceptable in our global way. We will use and conduct psychological economic operation by defining our have's and have not's first. We already discussed our have's, combining machines have but there is something on which we human are far superior than machines, which only we have and machine's can't even acquire it. These extraordinary superpowers of humans are very valuable in the field of economics. And these exceptional powers are;

a} Empathy;

Machines have the infinite power of cognitive psychology which make them ocean of knowledge and intelligence, completely incompatible to humans. But we humans also have cognitive empathy which machines don't have in their programming.

Cognitive empathy plays a significant role in society especially when it comes to economics where you have to deal with people not just the data's. In consumer market cognitive empathy gets applied, it can't be just solved by the previous data of consumer's analysis and draw future tactical. If you are handling the consumer market only on the basis of cognitive psychology and behavioral economics, your assumptions definitely fall astray.

b} leadership Psychology;

The beauty of leadership glows on unconditional and unfavorable environment. As for machines, empowered with all the leadership qualities even can't match with Man's leadership psychology. On the platform of economy where we have so many major concerns, the biggest concern we always face is the Time.

The loop of time with changing patterns of worldly things requires a numbers of orientations among changing knowledge of matters, economic patterns and organization's shifting paradigm. Machines are fueled with approximate data at particular interval of time, but the leadership of machines loses its dominance with speedily growing and advancing knowledge, system, knowledge and the regulations all are changing spontaneously with a fraction of seconds. These shifting parameters only demands the leadership of Man, because only man have the power of sense which transmit changing frequency to human mind for the process of new psychology which allow man to act accordingly and implement new version of ecosystem by which again machines update, therefore it's a clear fact that machines are dependent on human not all the time but in mean time.

Leadership is the subject which always fits on the hands of people only. The technicality which is spine of global world today is always operated and firmly oriented by people only.

c } Spirit ;

Too much of data's without spirit, Accountable data with the spirit. This is the fine line between humans and machines. The area of artificial intelligence has acquired all creativity except the spirit which only humans have. The power of spirit is solely responsible for will power, which consists of their own message and a will for change. There is no life without spirit, as for economy too without the spirit of will power, which promotes all market activities with their willingness and also generates willingness for others, can't be performed with machines and robots but with the human higher self, spirit.

These three exceptions of humans has maintained his presence and reserves his seat on global platform which still demands the leadership of people on global economic platform to refrain economy from paralysis.

2] Technicality;

Modernism is moving to advancement, where advancement has taken all places on the globe. With the advancement of time, economically we need to be more technical towards our organizational behavior. This technicality applies on resources, i.e. human resources, robotic resources, machinery resources, sustainable and non-sustainable energy resources and lastly economic resources.

In This age of advancement, Trade is the heart of economy which pumps Globalization and trading is the art of techniques.

The digital lens already provided the vision for future, which pictures high population and high demand of energies, for doing so we can't be just sitting on carbon emissions rather it's time for decarbonisation, combining all green energies with their sustainable powers to meet with basic need of coming generations and institutions. This will be technically promising factor for economy to deal with scarcity, resources; where machinery resources will be work on technical analysis and productive forces of society to enhance manpower and revenues too, which will be not functioning on non - renewable.

Meanwhile human resources will engaged in all decision making, managerial things, organizing and conducting or choosing pair of investments.

Technical investments will be mother of global economic world. In which we associate sustainable energy to Agriculture, non-sustainable to logistics and transportation, i.e., freight, air, railways and roads. And above both, there is mixed energy which demands more technical and practical approach of our minds because here we are combining both sustainable and non-sustainable at particular level with particular time for manufacturing and infrastructure unit. The application of mixed energy will be much technically analytical but fruitful, it also create balance in both the energies (sustainable and non-sustainable).

There is a huge demand of infrastructure or I say investments on infrastructure, these investments will failed to give better results if we use only one form of energy, as we all know the future of electricity, for whom battery power is working. Investments and incentives get higher with intentions, for any economy these things are power structures only when applied with technical approach.

By considering all upcoming shortages we have adopted technical move, and advantages of this technical move is that on the time of necessity, our pawn can be anyone of those, i.e. rooks, knights, bishop or queen.

3] The Invisible hand of Prices;

The evolution of price theory has been passed through various of stages but the ultimate end result of price is still invisible. In pre early times its just simply hand to hand, where consumer take his commodity from one hand and reciprocally pays from another hand. When series of revolutions started it had a great impact on pricing, Land was the first dictator of price with byproduct of wages.

$$P \propto W \qquad \text{(seen already)},$$

The amount of wages which entrepreneur has to pay to its labor gets added on price mechanism, this calculated amount of wages then decide selling price on market.

Landlords then entered in this race with their rents on land which again added with wages to decide price.

$$P = W + R,$$

The complexion of land, labor, wages and products needs an organization of proper capital amount to proceed and satisfy all promises. For that purpose money lenders lead that race with credit, which first given to entrepreneurs and returned with interest rates once they profited, this is the third phase of pricing system in combination with profit rate.

P = W + R + Pt

As world grows, people also grows up and when people are growing, economy get grow faster. Exact scenarios happened with price too, price and market price gets divided on the basis of demand and quantity demands;

X = Q + Qd + x'

At last, evolutionary theory of invisible pricing more puzzled by additionally charged services of; investments and speculations,

P = Iv + Fv + Sv + Is.

Price has always played a crucial role in any economy, meanwhile for global economy it needs to be more specific. Price seemed very under-rated in its theory of evolution, it doesn't has its own value, in each specific time for each specific factor its vector gets changed, not in a single time we find price as a deciding factor for any economic regulations; whether its demand, quantity demands or investments decisions. Perhaps there are many other factors which in combination decides how prices should operate, therefore we can conclude that there is not just one invisible hand but many invisible hands are involved in makings of market price.

We can't afford this pattern of price mechanism for global world, where we are not dealing with just one or two countries but the allies. By following this invisibility, we ended up with demolished global economy, far greater than Great Depression. Global Economy should have to adopt some reasonable methods which fit on both sides, i.e. developed and developing countries. Likewise pricing system of global economy also has to develop its own value by moving out from the rat race of credit cycles, and then take the dominating seat on global plane without any invisible hand but with so many helping hands by adopting one single method of Equity.

Equity will be the future of global world which will firstly apply for price mechanism, in where quantity demand and quantity supply should not decide what price would be but with the help of equity, price take its own form to operate on global platform by serving all countries equally, i.e. superpowers, third world or the one who is still battling with their hunger, shelter and cover.

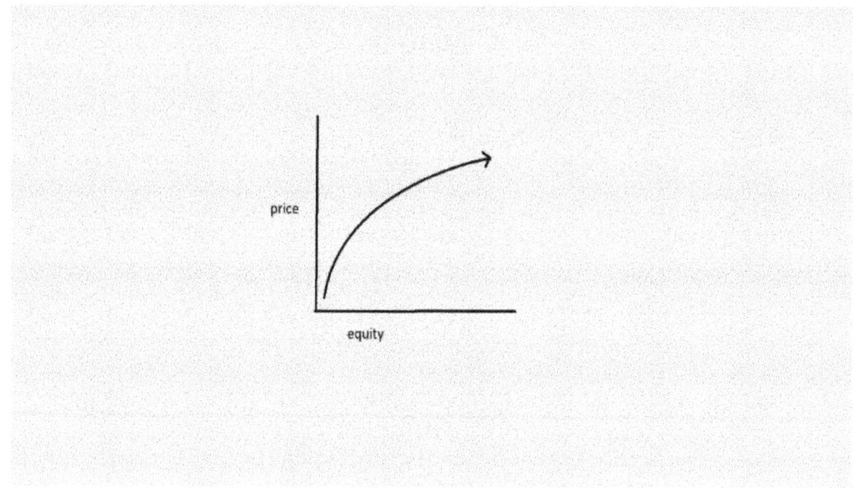

(Fig, 50; draws global pricing graph)

Price creates its own demand and supply on the basis of equity by allocative efficiency of economics. Both runs together also maintains the market price by controlling inflation with equity financing and circulation of capital at efficient rate without compounding interest rates per capital. This combination of equity and pricing led solid bricks of investments in several fields without any speculations, and also fruitfully revenue our unanswered question of surplus.

4] Demand and Supply;

Generally we are not allowed to follow actual demand and supply in our economy, whereas decisions are always form on the basis of quantities, economy is always a game of quantity, but as for global economy it should be quality not quantity.

The global graph for demand and supply based on quality has to be reform, where quantities match with equity, qualities allocate and efficiency shipped overseas.

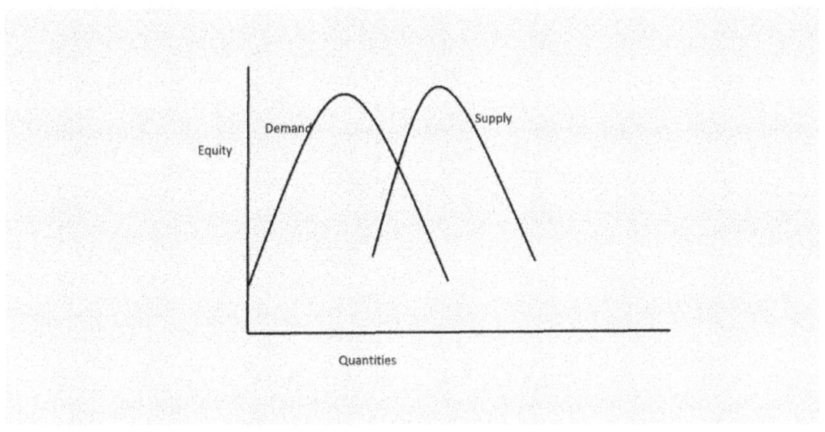

(Fig, 51; Global demand and supply)

The engine of equity in our global system of economy terrifically reshapes our quantity demand and supply curve with highlighting points;

a) Quantity demands finally chased by quantity supply at same level, with equity.

b) Both consumer and producer gain their needs at same time.

c) Quantities of both simultaneously rise in times of need and moves down after the completion of that specific need.

d) Because of equity, extravagancy of wants cuts down which is very prominent in dealing with scared resources for high population globally. It has a positive impact on business and market, where for each specific product demand has created not artificially by crediting forcible capitals but with efficient sincerity of applied capitals to perform businesses at certain time level. This demand is again get supplied by genuine trading, without exploitation of money in terms of high selling price.

$$E \uparrow - Qd \uparrow - Qs \uparrow \ldots\ldots phase\ 1$$

$$E \downarrow - Qd \downarrow - Qs \downarrow \cdots\ldots\ldots phase\ 2$$

(E- equity, Qd- quantity demand, Qs- quantity supply)

On phase 1, all are on rising state with the rise in price too. It also called as market rush, where everyone wants to try their hands on newly invented product and services to enjoy their benefits; benefits of demanded products, supply and financing of equitable funds.

On contrast, phase 2 appear after the satisfaction of market where all are on started falling apart with the thought of some new product and services.

But both of these phases doesn't take place suddenly as happens on business cycle, here it takes time to rise and decline because of equity and this constant time manages the price levels which is fuel of global world.

e) Price easily set on global dialogues with respect to new formed graph, it minimize negotiations and maximize surpluses.

5] Surplus;

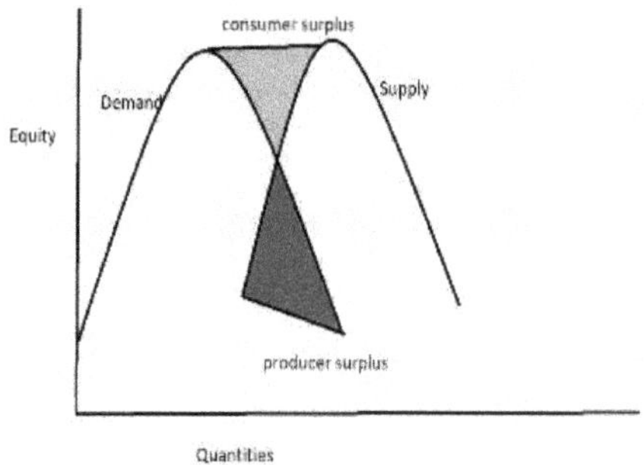

(Fig, 52; Global surplus)

According to the formulae of equity;

$$Equity(E) = Asset(A) + Liability(L)$$

This directly summarizes that removal of all kinds of liabilities gives us infinite number of assets in the form of surplus, which is additional with revenues. By simplifying this formula or subtracting liability we form equation of surplus.

$$Surplus(S) = Equity(E) + Asset(A)$$

The theory of Surplus starts with three phenomenal questions; which are, how to get money? Where to spend money? What to do with surplus? These questions not only solve surplus theory but again multiply it many times. The very first question, how to get money? Will be discussed and solved here. Whereas second and third will be answered in Blueprint, where we discuss economic organs.

To begin with first answer, as we already decided that on the global economy, regulators need to operate equity financing in market where everyone wants their individual gain on each side not specifically for anyone side. And for generating profit on each and every side there is only one way to begin with, which is corporate partnership.

When we get money from corporate partnership on the manner of equity not credit, it involves bilateral contract between two parties; where one is investor, another one is manager of that investment (this can be anyone either producer, manufacturer, trader or specialized manager itself) which has only one time contract till the completion of particular project, once invested project gets complete, now it's there call to renew their contract or dismissed that. During bilateral contract, disclosure of capital, profit, liabilities and dividends are must and already predefined with complete accountability, as far as profit is concerned is always shared among both parties as per their rate of capital invested in it, if there is only one investor then he has to set his managers rate of profit according to his effort and production. Whereas for loss only investor will bear it and his manager also not have any share or output from particular project. This type of financing module increase productivity, skilled employment, no concern of wages, only profit has major concern on both side, it will also exclude all kinds of Liabilities; no credit applies no interest rates and no interest rates applies no liabilities except profit generating initiative. And without any liability we only remain with assets (which comes from profit) and equity, their combinations lead us to our calculative formula of surplus.

$$Surplus(S) \;=\; Equity(E) \;+\; Asset(A)$$

This theory of surplus will only prove by corporate partnership. As we are moving for global leadership, where we will don't have to deal with one nation but the multiple of nations and their allies, to frame them on one global developed nation, which can only achieve not by per capita GDP of world but by per capita equity of world. Because if we go for per capita GDP, which financed by debts and taxes of governments we might fall globally.

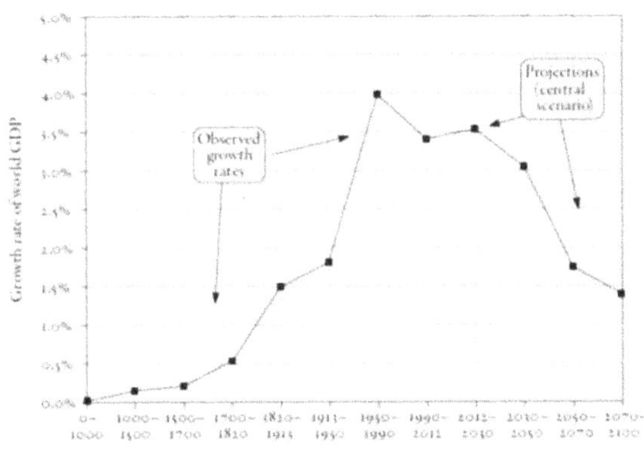

Sources and series: see piketty.pse.ens.fr capital21c. (75 76).

(Fig, 53; representing global GDP rate)

[86]*There are two main ways for a government to finance its expenses: taxes and debt. In general, taxation is by far preferable to debt in terms of justice and efficiency. The problem with debt is that it usually has to be repaid, so that debt financing is in the interest of those who have the means to lend to the government. (377)*

To recapitulate the argument thus far: I observed that an exceptional tax on capital is the best way to reduce a large public debt. This is by far the most transparent, just, and efficient method. Inflation is another possible option, however. Concretely, since a government bond is a nominal asset (that is, an asset whose price is set in advance and does not depend on inflation) rather than a real asset (whose price evolves in response to the economic situation, generally increasing at least as fast as inflation, as in the case of real estate and shares of stock), a small increase in the inflation rate is enough to significantly reduce the real value of the public debt. With an inflation rate of 5 percent a year rather than 2 percent, the real value of the public debt, expressed as a percentage of GDP, would be reduced by more than 15 percent (all other things equal)—a considerable amount.

Indeed, it is important to understand that without an exceptional tax on capital and without additional inflation, it may take several decades to get out from under a burden of public debt as large as that which currently exists in Europe.

On global platform we can't bear a fall, to avoid this fall we have to make right choices, A choice which transforms our society completely, these choices either make organs of economy alive or dead.

BLUEPRINT

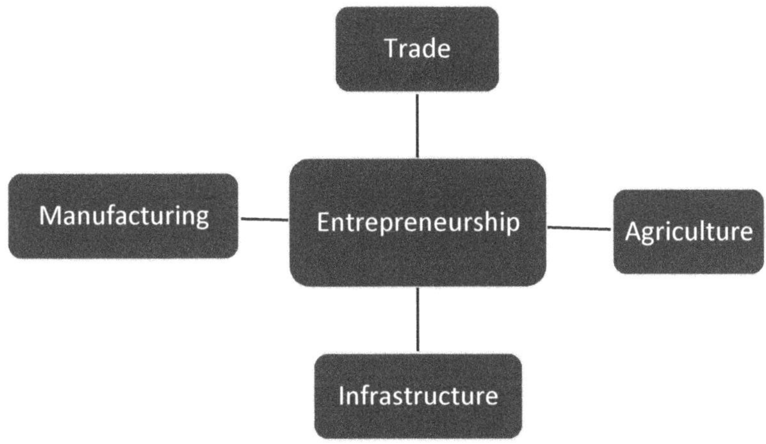

(Fig, 54; diagram of blueprint)

TRADE:

The heart of global economy is trade, as heart pumps all the blood throughout a body, likewise trade does, it flows money throughout the economy. Pumping is most prominent function of heart, in which blood with efficiency and on the mode of equitabilities reaches to several of organs. In global trading where convergence is necessary to speed-up, it requires combinations of capital from different investors to tie up in global project for production of surplus, because this is not a simple trading among nations anymore which easily sat on credit and preferred payments, it's a global play which needed win-win situation for all, either; consumer or producers, or; Regulators. For doing so, we categorized investor's _ manager portfolio along with accountable mechanism free from all errors.

Case Study 1;

We begin with global gateway for trade, UAE and its third largest trading partner India. India – UAE, our first case study of Blueprint.

Let's say India and UAE has contracted bilateral trade agreement, in which they both import and export their goods specifically, company A from India has decided to export food and beverages to company B in Dubai. Generally, trade in overseas performed when A offered his best product and services to B at his best pricing, for doing this A needs huge amount of capital too by following old path of mercantilism both A and B needs helping hands of bankers to borrow some money, A need it for selling and B needs it for purchasing, after the transaction from their individuals borrowed money both are reliable to pay their borrowings back, meanwhile A also waits for his LC's to release and while LC release to A, he witness completely changed future rate. This exact scenario happens with each country whether its third world or first world, they all have same pattern of trading which makes them trade deficit and horrifying balance of trade slows down the economy.

This outdated model can't be applied to technological equator of globe; we have to apply surplus generating formula as we are dealing with global economy.

$$E = m\, e2$$

When money is multiplied by equity and efficiency it will generate high surplus value economy globally not just for single country as a whole but for whole allied countries, for making that happen we have to apply leadership psychology with technical practice in which renewable and non-renewable resources get combined together for high performance.

E = in an economy of let's say India, company A enters in bilateral trading with company B of UAE,

m = the amount of capital allotted in a company's asset,

e2 = e – equity investments by different of investors to perform one contractual project at one time, in which profit is already mention in their agreement which will share among them as equity investors with bearing of loss too, in these investment projects numerous investors from all over globe can participate, these participation with transparency of profit evolves more and more investments. Foreign investors will not roll around loops of interest rates and future pricing, whatever profit margin initially get fixed they all gain exact value in any condition.

e – Efficiency in resources, whether it's primary, secondary or tertiary energy in combination with demand and supply of product and services, they all work upon efficiency. The amount demanded should not exceed supply and vice versa, with regard to energy resources most probably renewable energy totally applied in each sector with some conditional allotment of non-renewable.

A, invested all capital which is assigned by several investors for that one particular project, i.e. for exporting to B. one time disclosure agreement between investors and also with their appointed team leaders (also called managers) who is responsible for investors funds, by using these funds managers will be shipping all goods to overseas B, at their own security. As far as mode of payment is concerned, on the basis of equity formula, we apply payments from buyers after the delivery from A, when the

Boarding of goods to B occurred on his level of expected quality then he has to release all amounts at spot, no deferred payments apply here. Although it's a bilateral trade, it requires manager to return with imported goods to A, by doing this bilateral trading he not only just double company revenue but also utilizes resources of energy effectively. The shipped cargo will expecting to return after particular period, meanwhile manager complete all his previously decided imported goods from UAE to ship in India by clearing all tariff's and saving further separate importation and fuel.

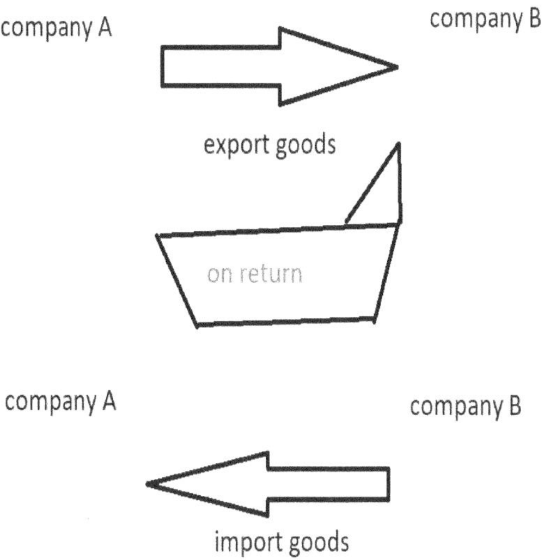

As we are heading to the age of uncertainty, hyper consumerism should replace by secure consumerism to achieve buyer's trust on global trading, payment methods play critical role in trading far critical than pricing, if payments should done at the end point after handling over to port, and all initial expenses manage by investor-managers cooperation, it will supernaturally enhance trading. Also by doing two-tier trading equity asset get high and trade deficit gets low with the maintenance of balance of trade, it's like win-win situation for both countries.

AGRICULTURE;

The most under-rated sector of economy which once solely had been brought revolution in history of civilization, but we refrain it and its value by exploiting its true nature with hazardous fertilizers, non-renewable energy resources and unwanted farmers debt.

Agriculture is like oxygen for an economy without oxygen there is no breath, no life; but in contrast hydrogen is continuously supplied in economy, in the form of financing, majorly agriculture financing is the rudest financing seen in society. Landing, laboring, waging, harvesting, unsustainable modern equipping all made farming costly and its outcome poor in favor of farmer itself, even in agriculture country like India alone we find approximately 170 million debts on farmers.

For evolution of agricultural sector, we need to begin with the same process occurred at the time of its revolution. It's all began in the Fertile Crescent almost about 12,000 years ago covering the Levant region of the Near East that includes the interior areas of present-day Turkey, Israel, Syria, Jordan, Lebanon, Iran, Iraq, Turkmenistan, and Asia Minor. Basically these regions are already empowered with plenty of sustainable resources, especially solar powers, canals, animal husbandry and rich amount of unoccupied `lands, if we transmit these qualitative natures with the technical collaboration of artificial intelligence it will going to produce more output than it had produced earlier, we just need to allocate correct system of financing with the sustainable power.

The mode of financing applied in agriculture development should be free from all kinds of mortgages, because here only asset which get mortgaged is land, and mortgaging land is like mortgaging natural resources which will never be fruitful to society. Instead of lending and mortgaging of lands there is much better option for global agricultural development and that is Forwards sale.

Forward financing on global market reduces risk and increases output, here; full advance payment get done to farmers without giving any loan and debts to them, just providing complete capital to crops by using renewable energy plants and technical help of machines lowers the human cost, whereby goods are delivered on future date decided by both the parties under quantity, time, and rate at the time of delivery according to their own needs, requirements, and specifications. This will not only cover all cost of capitals employed by investors but also provides them high margins in market, in case of default by farmers, i.e. non cultivated crops at time than farmers have to sell their lands to investors for specific period till they regain their invested amount on land, after getting their principal amount back they return farmers land on particular time, meanwhile farmers have no share in it, their shares are only dependent upon their cultivation of goods.

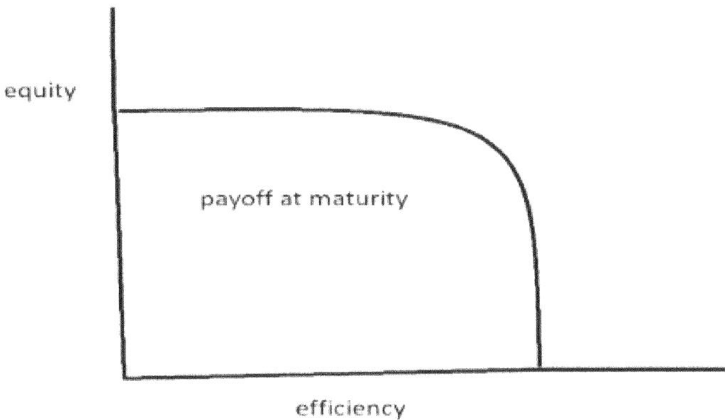

Efficiency and equity also plays major role here, the amount of equity invested at initial level has demand of exact efficiency of supply.

Manufacturing;

The big cheese since 1760 is no doubt manufacturing sector, in the era of AI also it holds stronger position the only difference will occur in this field is manpower replaced by machine power to make production at faster rate, reduce labor costs and increase GDP globally.

Europe has already enjoyed all luxuries of manufacturing outcomes at early industrial revolutions, as for future rising manufacturing hub will be seen on Vietnam, Mexico, India, Malaysia, Singapore.

Globalization of manufacturing required huge amount of investments with low risks, whereas generally investment method applied in manufacturing is either high lending or venture capitalist. Global world deny both because both of them don't have the capabilities to stand and run on global equator and meet manufacturing expectation of financing.

We have discussed widely on previous chapters about lending, credit, interest rates, their operations and damages. As far as vc's is concerned, its functioning is more complicating than interest with major involvement of risk management and revenue generation. Venture capital is purely based on Keynes's marginal efficiency pattern, where investor first venture capital amount to producers, say x; called it as seed funding, with seed funds producers began project of manufacturing as it goes towards maturity, financing of venture also mature with periodic time say, x1...x2...x3...so and so forth.

Investments like these complicate the entire profit generation on global platform, because it goes only in favors of investors to earn but also with uncertainty whether manufacturers gain will be low which directly lowers global goods output.

Future predicted manufacturing hubs like East Asia having two types of advantages, one is geographical and another one which will be helpful in financing manufacturing sector is population. These populations would act as an investors and fund managers by financing mutual funds on manufacturing.

Mutual funds always recommended for this sector but we used it in advance form, at very first there will be an agreement between investors and producers which will run and operate by fund manager appointed by group of investors, which have terms or clauses; 1. Pooled amount invested for 3-5 year period, which yearly accountable with balance sheet containing asset, liabilities and profit.

1. Profit percentage of investors are decided by their rate of investments, remaining's will go as revenue of company, subtracting all liability, only surplus of company remains with the surplus of investors.

2. Fund manager and investors also have contracted for this period, where manager has important role and according to his performance he get paid, he also have share in profit which was previously decided among them his profit percentage will be fixed and if he brings expected return then he get paid his percentage amount by investors, this pooling of capital are free from all wages of cost, it's purely based on contributions, the more effort full contribution has more fruitful profit. Manager also has right to invest in these capital, if he invested and manage this project than his revenue also has double share, first based on his investment and second by his work (already fixed rate of profit).

3. Mutual funds are majorly invested by ppp's- public –private partnership, where shares of companies issued and owned by people who productively want to participate and contribute with the help of stock market to enhance industries. By allowing ppp's mutual funds we have many Rothschild and lower level of inequality.

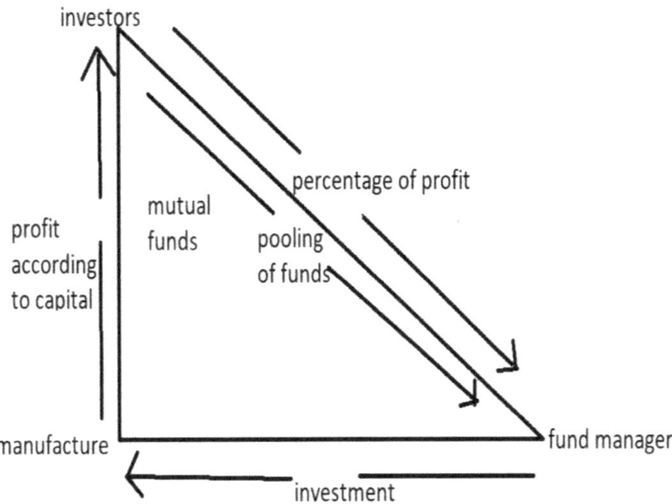

Efficiency in manufacturing matters a lot, as the amount of electricity required in manufacturing is very high so it has to be obtained from all renewable resources, especially by the wave power and battery technologies. Agriculture and manufacturing both sector has great opportunity for renewable resources other than trade and infrastructure.

Infrastructure;

Railroads started globalization of infrastructure by shaping all economic activities, i.e. trading, agriculture, and manufacturing together through a medium of vehicle which made revolution possible at bigger impact without that vehicle we haven't experienced these transformation all over perhaps without vehicle it just get restricted among some regions, but this vehicle of railroad infrastructure not just transported goods to all over surface but it enlighten unconsidered part of economy, which is infrastructure.

Although it's a long term project, it requires long term investment which can only be possible by partnership of public and private sector. Stock market, bond market, real estates are merely organized for these long term investment of capital; it rises up as project goes on with increase in share, it automatically increases value of share, at decline it loses its worth, there are operators or you can say their managers who knows it very well that when they have to pull liver up and down. People who invested in it earn accordingly but unfortunately this   method is also not promising and practice on trial and error platform.

Long term investment project should always have to operate by asset – backed financing. Asset has very definite role in long term financing because time is holding superiority when long term project implemented in economy.

Asset-backed financing is quite simple and promising for both parties, the central point of this financing is Time.

Also, it has one unique feature to interlink; infrastructure is usually a name of connectivity, same goes for financing of infrastructure too, it also interconnects economies together.

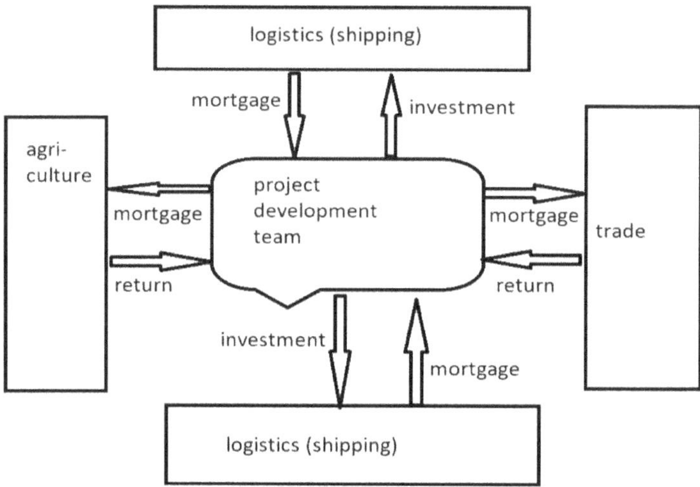

A project development team led whole interlinking process of investment by asset-backed financing where at very first team leaders invest on infrastructure project, i.e., logistics (shipping) or sustainable energy program for 5-10 years period of time, during this period development team also require a security for prolonged time being, as a security they take mortgage amount from producers of infrastructure for exact same period of investment. When the particular time completes they both backed their asset; team leader return mortgage amount and producers return investment value only (only principle amount), they are not participated in profit because their function is only to provide asset and on exchange of asset they reserve mortgage as security, meanwhile they also utilizes their security value by reinvesting on different of sectors like trade and agriculture with revenue generation in return from both till the completion of assigned time period.

Here, both equity and efficiency meets at right direction by expanding global production from all sides of economic sectors with minimum risk and speculations.

Entrepreneurship;

Entrepreneurship is a radius of economic cycle, when radius of any circle is calculated in a right way it will automatically form wealth generated economic cycle.

Each and every firm whether agriculture, trade, manufacturing or infrastructure, they all have one thing in common, they all need managers (team leaders) as a radius, without their central role none of the firm will have stability.

The upcoming world demands analytical leaders; entrepreneurs have a bigger place on coming times, this is a correct time to start initiatives towards microeconomic platform, which is a birthplace of entrepreneurs.

SME's, financial institutions, Equity market, sustainability, NGO's and trainings are the programs which uplift financial and economical literacy in society to reform entrepreneurs and leaders.

World is rapidly changing and everyone is chanting that employment is the biggest concern in future butt we forgot one thing that humans can't be replace from everywhere, AI can only replace those who are lacking in consciousness, artificial brain can be conscience but not more than nature. Our direct competitor is not human anymore it's even more than human in terms of intelligence and skills, the only way to survive in mass digitalization is by adapting entrepreneurship, because entrepreneurs are the only one who ride AI.

Summary of Book 2

- ✓ Old concepts of economy not applied in global world.
- ✓ New dynamics of principle set for global economic thought.
- ✓ Leadership is major demand in upcoming world.
- ✓ The most confusing chapters of economics get resolved in book2.
- ✓ Labor wage theory is not concerned anymore.
- ✓ Price theory also not troubling market rate.
- ✓ Global world should be debt free.
- ✓ Equity and efficiency reshape global economy.
- ✓ There is no presence of inequality; third world, second world or sole supreme power.
- ✓ Credit and interest market will crash whole global world.
- ✓ Energy and economics both have to be connected with balance.
- ✓ We already witnessed so many clashes, war, battles and collapses but this time if it will happen than it will be mass destruction.
- ✓ Humanity only has two choices; 1. Go with the worldly pattern or 2. Create a balance system on earth.
- ✓ All efforts on energy-economic sector which I have discussed and provided solutions is only for remaking of social and just society with balance of economy and truthful policies, concluding all; create a balance system on Earth.

References

1. Richard Feynman (1970). The Feynman Lectures on Physics Vol I.

2. Klaus Schwab- Shaping the future of the fourth industrial revolution, pg;9.

3. Klaus Schwab- Shaping the future of the fourth industrial revolution, pg;9.

4. Klaus Schwab- Shaping the future of the fourth industrial revolution, pg;10.

5. Source; world economic forum also, Klaus Schwab- Shaping the future of the fourth industrial revolution, pg;10.

6. Gordon 2016. And Klaus Schwab- Shaping the future of the fourth industrial revolution, pg;11.

7. Peter N. Stearns – Industrial Revolution in World History pg;15.

8. Peter N. Stearns – Industrial Revolution in World History pg;109.

9. Peter N. Stearns – Industrial Revolution in World History pg;202.

10. Alessandro Roncaglia- A Brief History of Economic Thought, pg; 10-11.

11. History Of Economic Thought - Harry Landreth and David C. Colander, pg;33.

12. History Of Economic Thought - Harry Landreth and David C. Colander, pg;38.

13. History Of Economic Thought - Harry Landreth and David C. Colander, pg;64.

14. History Of Economic Thought - Harry Landreth and David C. Colander, pg; 47.

15. Tragedy and Hope- Caroll Quigley, pg 50.

16. Tragedy and Hope- Caroll Quigley, pg 50.

17. Tragedy and Hope- Caroll Quigley, pg;51.

18. Tragedy and Hope- Caroll Quigley, pg;51.

19. Tragedy and Hope- Caroll Quigley, pg; 51-52.

20. Smith, Adam (1776). *An Inquiry into the Nature and Causes of the Wealth of Nations.* Vol. II (1st ed.). London: W. Strahan & T. Cadell. p. 35.

21. Murray N.Rothbard, Classical Economics – pg; 7&8.

22. Smith, Adam (1776). An Inquiry into the Nature and Causes of the Wealth of Nations. Vol. II (1st ed.). London: W. Strahan & T. Cadell. p. 11.

23. Smith, Adam (1776). An Inquiry into the Nature and Causes of the Wealth of Nations. Vol. II (1st ed.). London: W. Strahan & T. Cadell. p. 64 - 65.

24. Smith, Adam (1776). An Inquiry into the Nature and Causes of the Wealth of Nations. Vol. II (1st ed.). London: W. Strahan & T. Cadell. p. 225 - 226.

25. Smith, Adam (1776). An Inquiry into the Nature and Causes of the Wealth of Nations. Vol. II (1st ed.). London: W. Strahan & T. Cadell. p. 80.

26. Smith, Adam (1776). An Inquiry into the Nature and Causes of the Wealth of Nations. Smith, Adam (1776). An Inquiry into the Nature and Causes of the Wealth of Nations. Vol. II (1st ed.). London: W. Strahan & T. Cadell. p. Vol. II (1st ed.). London: W. Strahan & T. Cadell. p. 450 – 451.

27. Smith, Adam (1776). An Inquiry into the Nature and Causes of the Wealth of Nations. Vol. II (1st ed.). London: W. Strahan & T. Cadell. p. 841.

28. Smith, Adam (1776). An Inquiry into the Nature and Causes of the Wealth of Nations. Vol. II (1st ed.). London: W. Strahan & T. Cadell. p. 80, 81, 82.

29. Smith, Adam (1776). An Inquiry into the Nature and Causes of the Wealth of Nations. Vol. II (1st ed.). London: W. Strahan & T. Cadell. p.71.

30. Smith, Adam (1776). An Inquiry into the Nature and Causes of the Wealth of Nations. Vol. II (1st ed.). London: W. Strahan & T. Cadell. p.74-79

31. Smith, Adam (1776). An Inquiry into the Nature and Causes of the Wealth of Nations. Vol. II (1st ed.). London: W. Strahan & T. Cadell. p.134,135,136.

32. Karl Marx, Das capital – 1; pg: 104.

33. Karl Marx, Das capital – 1; pg: 104.

34. Karl Marx, Das capital – 3; pg: 219.

35. Karl Marx, Das capital – 3; pg: 246.

36. Karl Marx, Das capital – 3; pg: 244.

37. Alfred Marshall, Principle of Economics; pg: 61.

38. Alfred Marshall, Principle of Economics; pg: 62.

39. Alfred Marshall, Principle of Economics; pg: 275.

40. Alfred Marshall, Principle of Economics; pg: 179.

41. Alfred Marshall, Principle of Economics; pg: 50.

42. Alfred Marshall, Principle of Economics; pg: 341.

43. Alfred Marshall, Principle of Economics; pg: 343.

44. John Maynard Keynes, The general theory of employment, interest and money; pg: History Of Economic Thought - Harry Landreth and David C. Colander, pg; 293.

45. Eichengreen, Barry (2019). *Globalizing Capital: A History of the International Monetary System* (3rd ed.). Princeton University Press. pp. 5–40. *The Cambridge Economic History of the Modern World: Volume 1: 1700 to 1870*, vol. 1, Cambridge University Press, pp. 438–467.

46. John Maynard Keynes, The general theory of employment, interest and money; pg: 298.

47. John Maynard Keynes, The general theory of employment, interest and money; pg: 294.

48. John Maynard Keynes, The general theory of employment, interest and money; pg: 294.

49. John Maynard Keynes, The general theory of employment, interest and money; pg: 309.

50. John Maynard Keynes, The general theory of employment, interest and money; pg: 9.

51. John Maynard Keynes, The general theory of employment, interest and money; pg: 260.

52. John Maynard Keynes, The general theory of employment, interest and money; pg: 136 – 137.

53. John Maynard Keynes, The general theory of employment, interest and money; pg: 144.

54. John Maynard Keynes, The general theory of employment, interest and money; pg: 135.

55. John Maynard Keynes, The general theory of employment, interest and money; pg: 165.

56. John Maynard Keynes, The general theory of employment, interest and money; pg: 175.

57. John Maynard Keynes, The general theory of employment, interest and money; pg: 167.

58. John Maynard Keynes, The general theory of employment, interest and money; pg: 166.

59. John Maynard Keynes, The general theory of employment, interest and money; pg: 82.

60. John Maynard Keynes, The general theory of employment, interest and money; pg: 83.

61. John Maynard Keynes, The general theory of employment, interest and money; pg: 441.

62. Eichengreen, Barry (2019). *Globalizing Capital: A History of the International Monetary System* (3rd ed.). Princeton University Press.

63. Ray Dalio, Principles for dealing with the changing world order; pg: 276 -277.

64. Ray Dalio, Principles for dealing with the changing world order; pg: 277 – 278.

65. Tragedy and Hope- Caroll Quigley, pg;55.

66. Ray Dalio, Principles for dealing with the changing world order; pg: 285.

67. Ray Dalio, Principles for dealing with the changing world order; pg: 286-287.

68. Tragedy and Hope- Caroll Quigley, pg; 73.

69. Ray Dalio, Principles for dealing with the changing world order; pg: 312 – 313.

70. Ray Dalio, Principles for dealing with the changing world order; pg: 320,321,322.

71. Ray Dalio, Principles for dealing with the changing world order; pg: 322,323.

72. Ray Dalio, Principles for dealing with the changing world order; pg: 323-324.

73. Ray Dalio, Principles for dealing with the changing world order; pg: 323.

74. Ray Dalio, Principles for dealing with the changing world order; pg: 324.

75. Ray Dalio, Principles for dealing with the changing world order; pg: 340.

76. Ray Dalio, Principles for dealing with the changing world order; pg: 342-343.

77. Ray Dalio, Principles for dealing with the changing world order; pg: 344-345.

78. Ray Dalio, Principles for dealing with the changing world order; pg: 351-352.

79. Ray Dalio, Principles for dealing with the changing world order; pg: 353-354.

80. Klaus Schwab- Shaping the future of the fourth industrial revolution, pg; 57-58. And population reference bureau 2017, World Bank and institute for health metrics and evaluation 2016, world economic forum 2016a, world resource institute 2014, global challenges foundation 2017.

81. Klaus Schwab- Shaping the future of the fourth industrial revolution, pg; 88-89.

82. Klaus Schwab- Shaping the future of the fourth industrial revolution, pg; 99,100,101.

83. American psychological association 2013.

84. Klaus Schwab- Shaping the future of the fourth industrial revolution, pg; 167.

85. Klaus Schwab- Shaping the future of the fourth industrial revolution, pg; 169.

86. Some pictures resources; getty images, wiki resources, and some already mentioned below pictures.

87. Capital - Thomas piketty, pg;377-380.